SUNBURST ON MARCONI GLACIER

SIERRA NEVADA DE SANTA MARTA, COLOMBIA

FLAMINGOES FLOCKING OVER BOLIVIA'S LAKE POOPO

VOLCANOES OVERLOOKING LAGUNA VERDE

A STAND OF PALMS ON SIERRA NEVADA'S SLOPES

LIFE NATURE LIBRARY
LIFE SCIENCE LIBRARY
YOUNG READERS LIBRARY
GREAT AGES OF MAN
FOODS OF THE WORLD
TIME-LIFE LIBRARY OF ART
LIFE LIBRARY OF PHOTOGRAPHY
THE EMERGENCE OF MAN
THE OLD WEST
THE ART OF SEWING
HUMAN BEHAVIOUR
THE GREAT CITIES
THE TIME-LIFE ENCYCLOPAEDIA OF
GARDENING

THE ANDES

THE WORLD'S WILD PLACES/TIME-LIFE BOOKS/AMSTERDAM

BY TONY MORRISON
AND THE EDITORS OF TIME-LIFE BOOKS

THE WORLD'S WILD PLACES

European Editor: Dale Brown
Editorial Staff for *The Andes*:
Deputy Editor: Simon Rigge
Picture Editor: Pamela Marke
Design Consultant: Louis Klein
Staff Writers:
Michael Brown, Windsor Chorlton, Dan Freeman,
Heather Sherlock, Deborah Thompson
Art Director: Graham Davis
Designer: Joyce Mason
Picture Researcher: Karin Pearce
Picture Assistant: Cathy Doxat-Pratt
Editorial Assistants:
Ellen Brush, Elizabeth Loving, Jackie Matthews
Copy Staff: Julia West

Consultants
Botany: Christopher Grey-Wilson,
Phyllis Edwards
Geology: Dr. Peter Stubbs
Herpetology: David Ball
Ichthyology: Alwyne Wheeler
Meteorology: Lt. Cdr. Bruce Doxat Pratt
Invertebrates: Michael Tweedie
Ornithology: Dr. P. J. K. Burton
Zoology: Dr. P. J. K. Burton

The captions and text of the picture essays were written
by the staff of Time-Life Books.

Published by Time-Life International (Nederland) B.V.
5, Ottho Heldringstraat, Amsterdam 18

The Author: Tony Morrison is a zoologist, film-maker and writer. He has explored remote parts of South America, especially the deserts and forests of Bolivia and Peru. In preparation for this book, he toured the entire Andean range. He is also the author of *Land Above the Clouds*, a book about Andean wildlife.

Special Consultants: Peter Francis, a geologist, has travelled widely in the Andes in the course of his research into volcanoes and the chemistry of volcanic rocks. He is currently Lecturer in Earth Sciences at the Open University.

John Hawkes is Professor of Botany at the University of Birmingham. He has collected plants in most parts of the Andes and has a special interest in the botany of the high altitude regions.

Roger Perry is a zoologist and former director of the Charles Darwin Research Station in the Galápagos Islands. Since 1957 he has made nine extended journeys to the Andes. His book, *Patagonia, Windswept Land of the South*, was published in 1974.

The Cover: In the snow-capped Ecuadorean Andes, the peak of Sangay stands almost perpetually submerged beneath dense cloud. Difficult to reach due to the soggy jungle and ash on its slopes, this volcano stands over 17,000 feet high. It has been intermittently active for the 400 years it has been known.

Contents

Backbone of a Continent

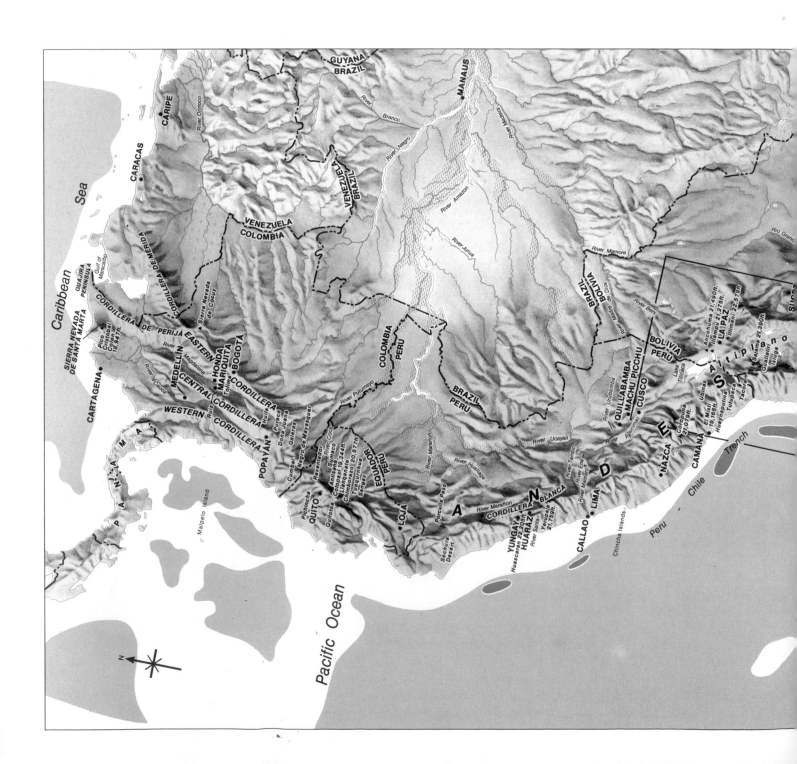

The Andes mountain chain—the longest continuous range in the world—extends 4,500 miles along the western edge of South America (green rectangle on the left). This young mountain system, which is studded with active volcanoes (red asterisks on the map below) and dormant ones (black asterisks), blocks rain-laden winds from the Atlantic and the Pacific, and creates two major arid areas (mustard), which receive less than ten inches of rain a year. The high plateaux of the cordilleras (green) get between ten and 40 inches of rain a year, while the eastern slopes of the central Andes and the west coast of southern Chile are drenched with more than 40 inches annually (blue-grey).

Off the Pacific coast, the Peru–Chile Trench is deeper than the Andes are high—over 20,000 feet (dark grey). The light grey zone runs from 10,000 to 20,000 feet deep; the white zone to 10,000 feet. The boxed-in area is enlarged on page 67.

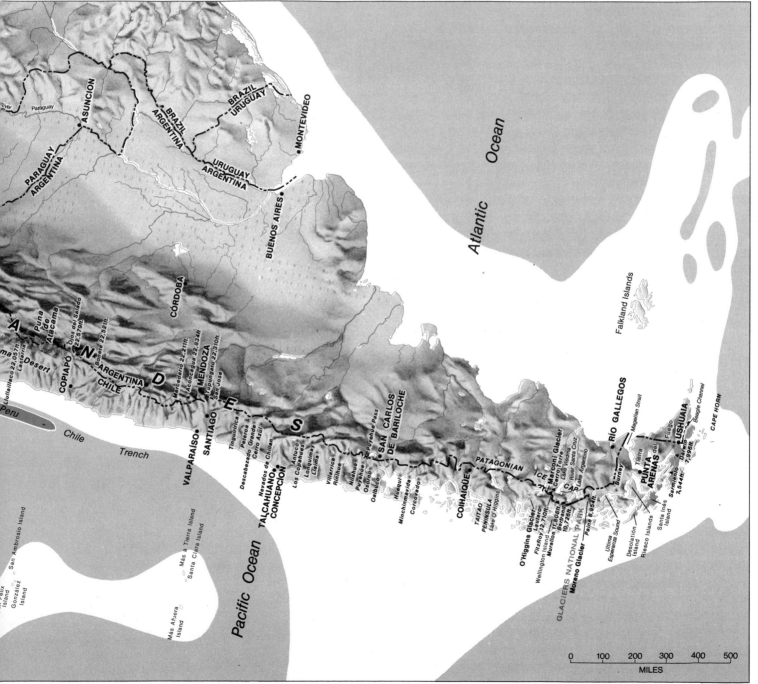

1/ Into the Mountains

Never had I seen mountains like these, and I was crushed by the grandeur—speechless with the overpowering wonder of it.

COL. P. H. FAWCETT/ *EXPLORATION FAWCETT*

The Andes are great humblers of men, and even of other mountains. They stretch the length of South America, forming a crenellated wall 4,500 miles long, draped at the northern end with vegetation, and at the southern end with ice and snow. Only the Himalayas and Pamirs of Asia boast peaks that are higher. Mount McKinley, at 20,300 feet, is the tallest mountain in North America; more than a dozen Andean peaks are taller. And down the length of this range, and on its slopes, lie wild regions of snow, ice and fire, of dripping jungle and seared desert, of cloud cover and merciless sun, of intense heat and killing cold, of warm moist air and air so thin that breathing is painful. Where, I wondered, studying my maps of South America, should I start? I made up my mind: I would begin where, for all intents and purposes, the Andes themselves begin, in northern Colombia, close to the shores of the Caribbean, and journey down them, travelling by car, truck, aeroplane and on foot, to where they end within 700 miles of Antarctica.

Thus it was that I and my wife Marion—whom I had met in South America several years before—found ourselves on the humid shore of the Caribbean, looking ahead into the mist in an anxious search for the mountains that we had come to see. It was only 6 a.m. and just light. We were on a narrow beach just outside the village of Palomino. Waves broke behind us and spread out in a lace of foam on the white sand. The air had that refreshing chill that only a cool morning in the tropics can bring. From the forest close by there came the grating shriek of a parrot and I could see small lizards scuttling in the dry growth along the edge of the beach. Two men came by on horses.

"Buenos días, buenos días," they said in friendly greeting, and then the older one, who had seen me craning my neck in the direction of the mountains, announced with the confidence of a native: "You'll not see the snows today. The clouds won't lift very far. The rainy season is nearly here." And with that, he and his companion trotted off.

In disappointment I turned again towards the Sierra Nevada de Santa Marta, as this group of Andean mountains, the highest in Colombia, is called. I hoped the man was wrong. He was—well, almost. I saw the mist disperse slightly to reveal a patch of snow. It was the flank of one of the peaks and it seemed remarkably close. I called Marion, who came running, carrying a camera. We waited for the view to clear, but just as quickly as the clouds had parted they closed in again and the Sierra Nevada de Santa Marta vanished.

Perhaps that was as it should have been. The Andes have never yielded easily to those who would know them. *Conquistadores* toiled into their heights in search of precious metals, and it was only within the last two centuries that men like the Prussian aristocrat, Alexander von Humboldt, and the young English naturalist, Charles Darwin, launched scientific investigations of the Andes and their exotic plants and animals. Standing on that beach, staring at the cloud-shrouded sky that hid the mountains, I felt some misgivings. Were we being over-ambitious? In the three months that lay ahead we hoped to sample as much of the scenery, climate and wildlife as the Andes have to offer. Would there be time enough for it all?

Too easily I could see in my mind's eye the maps of the Andes that I had studied when I laid our plans; and I reviewed mentally all the countries—and all the miles—that lay along our route. The Andes cut through seven countries. First there is Venezuela to the east, then Colombia, with three fingers of the range—the Eastern, Central and Western Cordilleras or chains—converging around Pasto near the Ecuadorean border. To the south is Ecuador, a land of volcanoes. From high on the slopes of Cotopaxi, itself a volcano, on a clear day a mountain climber can see 30 or so volcanic peaks all within the small compass of a hundred miles. South of Ecuador is Peru; here the Andes rise abruptly from the coastal desert and fall as dramatically in a torrent of vegetation to the lush world of the Amazon basin on the eastern side. At one point here the Andes are only 350 miles wide, and yet across that short distance, an amazing range of climatic conditions prevail, with a fascinating collection of plants and animals. Still farther south and eastwards in Bolivia, the Andes broaden to their widest point, some 500 miles across. And cradled in the depression between the Western Cordillera and the high peaks in the east is the plateau of the Altiplano which runs south for 500 miles.

Where they extend from Bolivia to Cape Horn, the mountains form the boundary between Chile and Argentina. During the Ice Ages an immense ice shield covered the southern Andes, and so heavily did it weigh on the narrow tip of South America that, when the ice retreated, the sea flowed in among the western mountains. Today the fjords of southern Chile provide some of the world's most desolate scenery: black peaks hung with cloud, drenched by rain and lashed by winds that often reach over 100 m.p.h. Of the ancient ice-cap there remain two large segments on the mainland and a tiny section on the archipelago of Tierra del Fuego, the southernmost extension of the Andes. In the cold of these realms the snow-line descends as low as 2,300 feet above the sea, and many glaciers grind down from the mountains right into the Pacific. I had timed my trip so as to arrive there in the depths of winter.

Where Marion and I stood, 4,500 miles to the north, the weather was different, and yet we could hope that in scaling the slopes of the Sierra Nevada de Santa Marta, we would have a preview of many of the climatic zones that occur along the north-south sweep of the Andes. We knew that for every thousand feet we would experience a 4° F. drop in temperature. We knew that the changes in temperature, combined with a variety of other factors including rainfall, produce broad bands of characteristic vegetation, and therefore we could also hope to see in capsule form some of the environments through which we would pass on our three-month journey. From sea level to approximately 5,000 feet lies the tropical zone. Merely by climbing between 5,500 and perhaps 8,500 feet, we could cross the subtropical zone, and another thousand feet would take us through the temperate zone to the so-called *páramo* zone that stretches to the snowline at 16,000 feet. From the *páramo* we expected to have a glimpse of the highest peaks.

Full of anticipation we started next morning for the mountains, driving along the coast from Palomino to Santa Marta, with Ariel Martínez, a Colombian agronomist, at the wheel. The land around Santa Marta was arid and the vegetation consisted mainly of tough, spiny plants that could withstand the fierce heat. The mountains reared almost directly behind Santa Marta, and so wild are they still, 400 years

At the northern end of the Andes, Colombia's Sierra Nevada de Santa Marta soars more than 18,000 feet above the Caribbean. Its cloaking vegetation, partly hidden behind cloud and mist, changes dramatically with increasing altitude. Between the thick rain forest on the lower slopes and the snows at the summit are a series of environments that duplicate many of those found down the entire 4,500-mile length of the Andean mountain chain.

after the Spanish first laid eyes on them, that only two roads wind part way up them. Beyond these, only mule trails straggle to the higher levels.

Our goal was to reach a mountain *cabaña* of the Colombian Forestry Service before nightfall. There we would find pack mules to carry our gear into the virgin forest that still covers much of the sierra. For a mile or two we drove along the main highway, and then we turned off to the right, taking a dusty road that ran beside a small stream. Very quickly the ascent began; the gradient seemed as steep as the line on a graph showing price inflation. We crossed a small spur that must have been a thousand feet above sea level. Although the land was as dry as it had been below, the vegetation was already greener, reflecting the favourable effect of the mist that often clings to the slopes at this altitude.

We pushed on through Minca—a small village of tin-roofed houses clustered around a police station and a bar—and soon found ourselves in a region of trees, taller than any we had seen at sea level. I asked Ariel to stop the car so I could get out and do a little exploring on foot. The road had been carved into the hillside, and the bank above was steep. I had been warned of snakes in the Sierra Nevada, and I made as much noise as possible to frighten off any poisonous specimens that might be lurking near by, knowing that in doing so I would also scare away other forms of wildlife. I took care as well to use only handholds that I could peer into, and I avoided reaching out to grab inviting looking ledges, lest they be the sunny resting places of venomous fer-de-lances and coral snakes. I saw how wise I had been to make so much noise when an 18-inch-long coral snake slithered into the grass ahead of me.

Reaching more level ground, I discovered a forest, mainly secondary growth that had sprung up after the primary stand had been cleared for a small coffee plantation or *finca*. Among the new growth were a few straggly coffee trees, and soon I came upon an ancient avocado. Before returning to the car, I stood a moment to watch a long line of ants moving in single file down a trunk and across the ground. They were leaf-cutting ants, and they were carrying neatly cut pieces of foliage to their subterranean galleries.

We resumed our drive and after negotiating a series of tight hairpin bends we rounded a curve and entered the lower reaches of the mountain rain forest of the subtropical zone. We welcomed the dampness of the air. It was refreshing after our tiring drive up the dusty mountainside. All around us the air was filled with hovering and darting hummingbirds, their wings beating so fast as to be almost invisible, although we could hear the vibrations. In the Andes, hummingbirds are found almost everywhere—from the Caribbean to Tierra del Fuego, and from tropical forest all the way up the mountains to the snow-line. If I needed a reminder that South America is "the bird continent *par excellence*", here it was. In Colombia alone there are more than 1,500 species of birds, more than half the total found throughout South America and twice as many as in the United States. What makes Colombia such good bird country is the equatorial warmth, the prolific plant life and the insects that the vegetation harbours or supports. The diet of hummingbirds consists largely of flower secretions and small, soft-bodied bugs. Their activity demands that they feed frequently, and since cold limits their food supply, it is no surprise that only one species lives in the chill southern Andes.

As we watched the hummingbirds, I could smell the forest—the rich fragrance of the many plants. I was sufficiently drawn by one sweet perfume to get out of the car and try to trace it to its source. But try as I did, I could not locate the flower. I even crushed some of the light green leaves of plants around the car to see whether it was they that were responsible for the scent but without success.

Advancing up the slope, we found that a cloud had formed beneath us, and we had occasional glimpses through its openings of the warm lowlands of the river Magdalena. The trail became gradually wet—dampened not by rain but by droplets of moisture from the cloud. The plants were wreathed in mist and the trees stood shadowy in it. The trail grew worse and Ariel had to drive slowly. We had almost a thousand feet to go before reaching the night-time shelter of the Forestry Service's *cabaña*, but already dusk had overtaken us. Ariel flicked a switch to turn on the headlamps. Nothing happened. Another click and still no lights. Luckily Ariel had been over the trail many times, and by leaning out the window he could see just enough of the road to continue.

We did not discover until some days later that there was a terrifying drop on Ariel's side of the car. On our way down I happened to look out the window, and was startled to see that what I had assumed on the way up to be a wall of vegetation had in fact been darkness, concealing an abyss. Even though we were happily unaware of the danger of the drive, those last moments were tense. The only light I could see was the glow from the stars above and the occasional flash of a firefly. And then, reassuringly, the lights of the *cabaña* glittered up ahead in the darkness.

The cabin lay at 7,200 feet, and when Marion and I climbed out of the car, stiff from the long drive, we were stung by the cold night air. I dug a sweater from my rucksack and pulled it on. After supper we went directly to bed for we wanted to be up with the dawn. Waking we found a perfect morning. But still we could not see the snowfields on the slopes above; they were hidden by a steep *cuchilla*, or knife-edge ridge, that stood blocking the valleys leading up to the snow-line.

Our mules were waiting for us in a grassy paddock. As soon as we had breakfasted and loaded the animals, Marion, Ariel and I set out. A cloud slowly formed and wrapped us in its warmth; it could not have been very thick, for the glare from the sunlight penetrating it hurt our eyes when we looked up—and yet it was to stay with us for almost the rest of the day. The mules moved quietly on the black, fibrous soil of the forest floor. Ariel said that the valley lay far below to our right but because of the cloud we could not see the drop—indeed, we could see only a few yards around us. From ahead came the crashing sound of a waterfall but just how far away it was, I could not gauge: even a tiny stream trickling over rocks can seem like a torrent in the rain forest where all sounds are somehow amplified and given resonance under the dense canopy of branches and leaves.

As we moved higher the tangle of bushes and trees grew thicker. The path dipped slightly to curve around the neck of the valley. Here a stream flowed noisily between moss-covered boulders and the mules splashed through the clear water without slipping. Now the path led upwards towards another *cuchilla*. In such steep country, progress is invariably slow. Trails in Andean forests seldom zig-zag up the mountainsides but usually follow the contours instead. A journey between two places that are quite close on the map can often take several days to complete.

At the end of another valley the vegetation changed once more and we entered the gloom of the high mountain rain forest. The trees towered overhead now. The long straight trunks, some of which I estimated to be 60 feet high, rose into the enveloping cloud. I strode over to one, a podocarp, distantly related to the pines, and tapped the huge trunk the way a landlubber knocks the wood of a boat to see how solid the vessel is. The tree was at least four feet across, smooth and utterly immoveable. And yet I knew that for all its apparent strength it was as vulnerable as a sunflower in the wind. Such trees can easily topple. Once I had been in a similar rain forest on the frontier between Bolivia and Peru. It was late afternoon and a squall blew up and lashed the trees. The storm lasted nearly an hour, and I could hear the forest titans crashing to the ground around camp. The next day the air still reverberated with the sound of broken branches that had been caught in the

A snow-mantled ridge of the Sierra Nevada de Santa Marta snakes eastwards from Colombia towards the Venezuelan border. In spite of its tropical location, the sierra is permanently snow-covered above about 16,000 feet. But as the Andes run southwards the snowline edges lower, dipping to a mere 2,300 feet in Tierra del Fuego.

branches of still standing trees and that were now breaking free and joining the giants on the forest floor.

I moved a few yards into the forest. The ground was soft, although not wet—the rains had yet to come. In this muffled world the stillness was broken by only a few sharp insect noises and the loud, strident croaks of tiny frogs. Then when I least expected it, a deep roaring swept eerily through the trees; it was a sepulchral note, and Ariel must have seen me start for he was quick to say, "Monos". Seeing I did not understand, he added, "Mono cotudo"—red howler monkey, the second largest wild primate in South America. Normally I would have recognized the howler's cry, but I had not expected to encounter the species so high up. In fact there must have been a group of them below, in a narrow, densely wooded gorge. We continued up the mountain with the mules plodding on ahead. On one side of the trail a tree had fallen, opening up a space in the canopy and letting enough light through for rampant growth to have occurred on the exposed site.

We trudged on for a while and then paused so that I might sweep the canopy with my binoculars for signs of animal life. Normally one would expect to find toucanets and parakeets among the tough, leathery leaves of the upper storey, or at least flitting butterflies. Although I scanned the vegetation carefully I saw nothing, not even a hint of movement in the leaves to give away a hiding animal. Instead I picked out epiphytes clinging to the trunks and branches—orchids, ferns, bromeliads, with their whorls of leaves holding reservoirs of water in which whole communities of animals can exist. I reflected a moment on what conditions were like 4,500 miles to the south at the same altitude; there the ice-fractured rock would be swept by 100-mile-an-hour winds, and below, instead of the jungle canopy, would be the congealed mass of the ice-cap.

By the time we reached the clearing where we were to camp for the night, the warm cloud that had hovered over us and at times surrounded us during the day was suffused with the pink and gold of the setting sun. As the light dimmed and the air began to cool rapidly, the scene quickly changed—a bit of theatre in the mountains. Balmy masses of cloud streamed past us on currents of swiftly rising warm air. The mist lifted from the tree-covered slopes, while still more cloud rushed up from the valley and drained over the *cuchilla* ahead. Suddenly the coastline appeared far below in pin-sharp detail. The cloud, which had taken hours to form in the morning, had disappeared in minutes. Only a tiny puff was left; it could have been the smoke from an exploding firework.

We relaxed and watched night cover the mountains. Ariel made coffee, a black brew, which we drank with heaps of sugar while sitting around a bright fire. Our tent was a simple one—a long sheet of green plastic draped over a crude frame of branches. In buying the plastic, Marion and I had hoped to make our camp blend with the environment, but when we awoke the next morning, we had a surprise. The plastic was all too green against the subtler shading of the rain forest.

The air was wonderfully clear, and we watched a small ship, way out at sea, leaving a silvery wake in the water. A heat shimmer formed above the brown scrub of the tropical zone which we had explored two days earlier. The brown turned perceptibly green where the mountains began and then the green darkened and ran through various shades all the way up the slope to our look-out just below the cold, drier *páramo* zone. All around us were lichen-covered bushes.

The morning seemed full of promise; at last we would reach the top of the *cuchilla* still obscuring the snow-covered peaks. Without waiting for breakfast, I climbed towards the top of it. The plants underfoot were stubby and thick-leaved, with a grey-green tinge that I soon recognized as the dominant colour of these heights. The sun was heating the ground, and as though in response to the warmth, insects were already making noises. Once a bird called out. I reached the top of the *cuchilla*, and looking up, beheld the gleaming snow.

I was still some miles from it; a wide valley lay between me and the mountain tops, but I could see clearly the line of dazzling white peaks. And there in all its supremacy, rising above the others into a sky as blue as a freshly-opening morning glory, was Mount Cristóbal Colón—at 18,947 feet the fourth highest mountain in the Americas north of the Equator. This magnificent panorama whetted my appetite for more, and I was left with the certain feeling that for all the trouble that might lie ahead for us, our journey along the Andes would be worth it.

The Restless Skies

In weather satellite photographs of South America, the Andes may appear as an unbroken line of cloud curving for 4,500 miles down the west coast of the continent, from the Caribbean to the Southern Ocean. Viewed from the ground, though, the cloud cover is seen to have many permutations, and these create the characteristic moods of the Andes—from the balmy mist of a Caribbean dawn and the haze of the coastal desert to the cold sheen of frozen fog over Patagonia and the billowing clouds that hide the highest peaks.

Much of the cloud cover is created by the Andes themselves. Reaching nearly four miles into the sky, they obstruct approaching winds, forcing them to rise into cooler air where moisture is condensed into cloud, often rain-bearing cumulo-nimbus.

Along their great length, the Andes intercept two rain-bearing air masses in this way. The cordilleras in the north divert the easterly trade winds, blowing in over the Amazon from the Atlantic, and the cordilleras in the south take the brunt of the westerlies or roaring forties, coming in from the Pacific. Both winds have the same dramatic effect on climate, drenching the windward slopes and leaving a desert in the "rain shadow" of the lee.

Within these wind systems, individual peaks, like Aconcagua in Argentina, the highest in the Americas, create their own micro-climates. Isolated above the surrounding landscape, they are frozen longer than their neighbours and more frequently shrouded in high clouds.

Many of the lower Andean slopes are regularly draped in fine stratus cloud. In the coastal Andes of Colombia, warm damp air often drifts in from the Caribbean overnight and is chilled by contact with the cool valleys. The mist that forms (right) rises in the warmth of the morning sun as stratus cloud, evaporating by noon to reveal a vivid blue sky.

Farther south, a pall of stratus casts a winter-long shadow over the coastal foothills of the Atacama desert. It forms because the lower layers of moist Pacific air are cooled as they pass northwards over the Peru-Chile current. It drifts up over the coast to about 2,000 feet where a stable inversion level of warmer air inhibits it from rising farther.

In the far south, gales continually sweep fog, cloud and powder snow across the Patagonian ice-cap, in frigid contrast to the mellow warmth of the northern Andes.

Dawn in the valleys of Colombia's Sierra Nevada de Santa Marta is a period of dramatic change. Mist that has formed low in the foothills during the cool night is ruffled by turbulence created by the warm morning sun. The mist rises as stratus cloud to about 3,000 feet, with wisps as high as 8,000 feet.

A curtain of mist lends a deceptive aura of dampness to the drought-stricken desert of coastal Peru, while nurturing drought-adapted vegetation such as this candelabra cactus. All winter the mist, formed when moisture in Pacific winds is cooled by the influence of the off-shore Peru-Chile current, drifts into the foothills. It then rises as stratus cloud to about 2,000 feet where an inversion level of warmer air creates the distinct line in the sky (left).

A shaft of lightning flashes to earth near Lake Poopó, high in the Altiplano of the Bolivian Andes. The discharge rips directly from billowing cumulo-nimbus cloud—whose lower layers bear a preponderance of negatively-charged water droplets—to the positively-charged earth. A deluge of rain is released from the huge cloud base.

In spring and in early summer on
windswept Mount Aconcagua, two sets
of weather conditions prevail at sunrise
(above) and sundown. Minutes after
sunrise, early morning turbulence
creates tufts of cumulus over the
surrounding mountains and valleys.
These gradually build into towering
storm clouds that bring with them
violent showers of rain or hail.
Throughout the day, jet-stream winds
blowing at up to 150 m.p.h. whip the
powder snow off the summit into a plume.

After sundown, although the jet-stream winds persist around Mount Aconcagua's summit, the clouds that have been building all day have subsided in the falling temperatures of the late afternoon. Only the cloud tops remain as a layer of mottled cirro-cumulus cloud that is propelled by the wind around the mountain in a crescent-shaped swath. Within the crescent, the plume of wind-blown snow is still visible, rising from the summit of the mountain.

Cloud, fog and powder ice shroud the gleaming surface of the Patagonian ice-cap in the far southern Andes. Thin strands of alto-stratus cloud (background) drift across the face of the early morning sun while fog swirls around the base of Mount Pirámida. What appears to be cloud in the foreground is in fact powder ice or spindrift swept up by gale-force winds.

2/ On the Trail of the Botanists

Two days after our ascent of the Sierra Nevada de Santa Marta, Marion and I travelled westwards along the flat, swampy Caribbean coast. We were following the direction taken by generations of Andean explorers since the Spanish sailed along this coastline at the beginning of the 16th Century. From here no mountains were visible and the first sign of the Andes proper was the muddy outflow of the river Magdalena, which stains the sea for several miles out with a heavy load of sediment. From this tell-tale evidence a thoughtful traveller might infer the presence of a great range of mountains somewhere in the interior of the continent.

They lie to the south, beyond the wide flood plain of the Magdalena. For a hundred miles inland, the country is monotonously flat, patched with lagoons and marshy areas. Much of it is covered with tropical jungle, sustained by the baking equatorial sun and a phenomenal rain-fall of more than 200 inches a year. To the west of the river, open grass-land appears as the ground rises towards the mountains, and to the east the flat lands are bounded by an arm of the Andes which reaches north-wards towards the Sierra Nevada. This stretch of jungle and savannah is the *tierra caliente* or hot country of Colombia, and the river journey through it is the traditional first leg of a trip into the northern Andes.

The starting point for such a journey has always been the port of Cartagena, about 60 miles beyond the Magdalena, where Marion and I arrived in the afternoon. Cartagena was the maritime capital of the

Spanish in South America. For three centuries every ship arriving in South America had to call at this fortified stronghold before going on to other destinations, in the same way that ships bound for Central America had to call at Puerto Bello and those bound for Mexico at Vera Cruz. In this manner the Spanish kept the key turned on their rich South American possessions. They also locked away from scientific investigation the magnificently varied flora and fauna of the Andes. Even at the end of the 18th Century, only a handful of outsiders had penetrated the interior of the continent.

One of the most significant breaches in this policy came in 1799, when the young Prussian aristocrat and naturalist, Alexander von Humboldt, was granted permission for an expedition to study anything he wished "for the progress of the sciences". Humboldt's excellent connections in Madrid won him an audience with Charles IV of Spain and enabled him to infect the king with his enthusiasm for the natural history of Spain's New World colonies. Within a few days of his presentation, he was given letters patent requesting all officials of the colonies to give him assistance. Two years after his departure from Europe, after exploring the lowlands of Venezuela, Humboldt, accompanied by the French botanist Aimé Bonpland, disembarked at Cartagena for the first unrestricted journey of scientific investigation made in the Andes by foreigners. The two men travelled through the northern Andes as far as Lima in Peru, pausing at various places on the way, especially Bogotá where they were welcomed by a Spanish priest, José Celestino Mutis, then the foremost botanist in the New World. In the course of their travels they assembled a mass of information that was later published in Humboldt's voluminous works on botany, zoology, geology, surveying and related subjects.

For my latest journey into the Andes, I decided to act on an old idea of mine and follow in the footsteps of Humboldt and Bonpland, travelling as far as the high, wet grasslands or *páramo*, south of Bogotá. The explorers had started by crossing the *tierra caliente* on the river Magdalena, toiling upstream for six sweltering, insect-ridden weeks, but I decided that we would not re-live this stage of their journey. Instead, Marion and I took seats on the afternoon flight direct from Cartagena to Bogotá, now the capital of Colombia. We were soon following the course of the Magdalena, curling 20,000 feet below us through the dense jungle of the *tierra caliente*. We passed over the small town of Honda, where the notorious cataracts had forced Humboldt and Bonpland to continue their journey on mules. Shortly afterwards, the aircraft turned east. There was slight turbulence as we flew into a cloud bank. Suddenly we

came out on the other side. I was not prepared for the effect. One moment we were over the Magdalena valley, the land far below; the next, the mountains had soared up and were close beneath us.

We had flown in between the spread fingers of the mountains, where the Andes separate into three distinct ranges—the Western, Central and Eastern Cordilleras—and were now over the Eastern Cordillera. A carpet of green clothed the mountainsides, and thick jungle choked the bottoms of the valleys. This rich growth of plants is the characteristic feature of the northern Andes, and it continues unbroken but for the snowy summits of the high peaks, right down through southern Colombia, where the cordilleras entwine into a single mass, and on through Ecuador. Only in Peru does the green begin to fade, giving way eventually to the cold, high deserts of the central Andes.

We edged down through patchy cloud to approach Bogotá and the ground fell away into the open expanse of the *sabana*, the dry lake bed at 8,600 feet on which the capital is built. Bogotá is a fast-growing colonnade of concrete aimed skywards, and as we came in to land I had my doubts whether I should find any trace of the early naturalists. One task I wished to fulfil before exploring the *páramo* was to see some of the botanical work done by the priest Mutis. When Humboldt and Bonpland arrived in Bogotá, on their journey of exploration, Mutis was already an old man of 70 and had largely completed his life's work, a comprehensive study of the flora of New Granada, as Colombia was then known. Ever since leaving Spain in 1760, he had been travelling and cataloguing, despatching descriptions and specimens of Andean plants to the royal botanical collections in Spain, and corresponding with the great Swedish botanist, Linnaeus, who had once been his teacher. Linnaeus had predicted, before any serious work was done in South America, that the continent would prove to contain many plants new to science, and Mutis confirmed this time and again by his field-work. To help handle the quantity of information he gathered, Mutis had assembled a team of draughtsmen and students, and I imagined that their output over 40 years must have been enormous.

My quest for Mutis's work at first seemed ill-fated. In spite of his importance, the botanist appeared to be almost unknown in the Colombian capital. I discovered the first trace of him quite by chance. On the day after my arrival, while I was exploring the old part of the city, I came upon a small square. Three sides were filled by a white-painted colonial house with red tiles and green doors set at intervals.

From the forest-choked valleys of the northern Andes rise the rumpled slopes of the upland wilderness, carpeted with low-growing plants.

Unexpectedly, beside one of these doors I found a plaque that read *Clausura de la Real Expedición Botánica*—Office of the Royal Botanical Expedition. This was the suitably grandiose title that the Spanish Crown had given to Mutis's 40-year survey of Colombian flora.

The door was locked and the place looked closed. But not far away I found a tiny bookshop open and went inside, determined to ask about Mutis. The only light came from two bulbs hanging beneath dust-covered plastic shades. The ceiling was high and every wall was lined with tiers of paper-bound books. A small, grave man came forward.

"Buenos tardes, señor, a sus ordenes." He adjusted the lapels of his neat grey suit.

I returned his greeting, and asked for any books on Mutis.

"Mutis, señor?" He rubbed his forefinger reflectively behind his ear. "Si, señor, but the books are very expensive."

He went to a bottom shelf and returned with a folio volume two inches thick, beautifully bound in soft, light brown leather. The lettering was in stamped gold leaf: *Flora de la Real Expedición Botánica del Nuevo Reino de Granada*. I opened it. The paper was heavy and smooth, and each page was interleaved with thin, crisp tissue. There were numerous plates. As I turned through them, here in the middle of the city, the vivid world of the green Andes came alive. Almost the whole volume was devoted to the passion flowers of Colombia, their climbing vines and slender tendrils drawn in fine detail by the artists Mutis had employed. The flowers were coloured in brilliant blues, deep crimsons and rich ochreous yellows.

"This book must be very old," I said.

"No, señor. It is new."

He went on to explain the strange story behind the book. When Mutis died in 1808, most of his papers and specimens were sent to Spain at royal command, and handed over to the *Museo de Ciencias Naturales de Hispania* in Madrid for safe keeping. There were 14 cases containing 5,190 drawings and 771 sketches of plants; one case of manuscripts, 48 cases of plant specimens and 41 cases of seeds, samples of wood and drawings of Andean animals. However, the onset of colonial revolutions in South America and the Napoleonic wars in Europe pushed scholarly pursuits into the background, and the materials gathered by Mutis were forgotten until well into the 20th Century. The first volume of his work was not published until December 1953, as part of a joint enterprise between Spain and Colombia. Several more have now been printed, but the projected total of more than 50 volumes will take many years to

Among the many varieties of passion flower described and illustrated in the *Flora de la Real Expedición Botánica del Nuevo Reino de Granada*, the life's work of José Celestino Mutis, was this *Passiflora mollisima*. The specimen originally used to describe this species was collected by the young naturalists Humboldt and Bonpland near Bogotá, while they were visiting Mutis.

prepare, creating the extraordinary possibility that first publication will be completed as much as two centuries after the author's death.

Humboldt and Bonpland spent two months with Mutis in and around Bogotá, and when they began the next stage of their journey southwards in September 1801, they were accompanied by Francisco José de Caldas, one of Mutis's most brilliant young disciples. Caldas was a native of Popayán, 200 miles south of Bogotá, and at Mutis's request he took the visitors to see the bleak open spaces of a *páramo* that lies in the mountains above the town. We planned to explore the same wild country, and Marion flew ahead to Popayán to make preparations for the expedition. Meanwhile, I travelled through the intervening deeply folded mountains by road, keeping close to the route taken by Humboldt.

From Bogotá I drove west, down into the Magdalena valley on whose slopes the naturalists had passed through a forest of wax palms, tree ferns, orchids, passion flowers and fuchsias. I went through the town of Mariquita, where Mutis lived for eight years and collected flowers and plants so numerous that it took 60 pages of his journal to describe the results of one field trip in 1783. From the Magdalena, I crossed the Cordillera Central along a spectacular route leading over deep ravines, narrow tumbling rivers and valleys thickly entangled with virgin forest, then drove south for 200 miles down the rich valley of the river Cauca until I reached Popayán.

Much of Colombia's temperate zone is intensively cultivated, and many small colonial towns of the 18th and 19th Centuries have grown into large industrial centres. But once you leave the excellent modern roads you soon find yourself on mule trails and among forests where the plant life observed by Mutis and later Humboldt is as riotous as ever. Many times, as I ascended or descended mountains, I passed through tropical, subtropical and temperate forest and back again within a few miles, and I saw quite a cross-section of Andean plants.

On the slopes of the Eastern Cordillera above the Magdalena, I entered a stretch of subtropical forest, "a marvellous stratum of life", as the Andean ornithologist, Frank Chapman, described it. "Every available foot of ground is claimed by trees and undergrowth, and every available inch of the trees is claimed by parasitic or epiphytic vegetation." Flowers glimmered dimly in the scented humidity. I caught the soft sound of insects and glimpsed the iridescent flash of butterfly wings. Among the trees I found a passion flower, its pale lavender, almost white petals shining like a star in the tangled undergrowth. The tendrils were long and tightly twisted, holding the plant to its neigh-bours as it climbed from the forest floor. The details of its construction were as sharply defined as in the pages of Mutis's *Flora*. I had to look twice to reassure myself I was seeing reality and not its image.

On my way up into the Cordillera Central I discovered some dense thickets of the giant South American bamboos known as *guaduas*. These plants were mentioned in the earliest Spanish chronicles since they were difficult to penetrate and made ideal hiding places for the native tribes when they mounted ambushes on the invaders. I also saw tall, elegant palms whose slender greyish trunks were topped with crowns of dark, shiny leaves. These were the wax palms mentioned by Humboldt. The exploitation of the waxy substance they secreted led to their widespread destruction, but now they are protected and have an established position as the "national tree" of Colombia.

In a belt of particularly thick, wet forest, I noticed some strange logs lying on a muddy bank. They were five or six inches in diameter, dark brown and rough on the outside, white and pithy in the middle. They had not been sawn but cut cleanly, apparently with a single swipe of a razor-sharp machete. The white, exposed surface was quite soft, but scattered across it were some mahogany-coloured patches, and when I pressed my thumbnail into these parts, I discovered that they were tough. To try to find the plants from which the logs had been cut, I pushed into the damp vegetation. In the cool darkness of the forest, I found a giant tree fern at least 15 feet high, maybe much more—I could not be sure because its base was hidden in the thick tangle of under-growth. There were more giant ferns farther into the forest, and I realized that these were the plants that had provided the logs. The trunk of a tree fern is closer in construction to a green plant stem than to the woody bole of a normal tree. A series of water-conducting cells run up the stem like veins, and when the fern is cut down, the exposed cells discolour and harden, forming the curious mahogany-like patches that had caught my attention.

Often in the silence of the forests I sensed the soft, quick vibrations of a hummingbird's wings. Bird species, like plant life, are concentrated in remarkable numbers in the subtropical zone, where the abundance of insects, flowers and seeds affords the richest of habitats. Well over 300 species of Andean birds are found at this subtropical level (roughly 5,500 to 8,500 feet), and more than half of these occur nowhere else. One of the most interesting is the oil bird, *Steatornis caripensis*, which was first described by Humboldt. The young oil birds spend an unusually

The symmetrical rosettes on the right are common among plants growing in the páramo zone of cold, wet grasslands, more than 12,000 feet up in the northern Andes. The low, compact shape of the plants reduces wind friction and curtails heat loss, and the overlapping leaves shield the vulnerable core. Even the flower of the Vaccinium, red like those of many bird-pollinated plants, has a rosette arrangement.

ERYNGIUM SPECIES

ESPELETIA SCHULTZII

YOUNG VACCINIUM LEAVES

RED VACCINIUM FLOWER

long time in the nest, during which they become inordinately fat, and it is from this feature that the species acquired its name: the Indian tribes used to "harvest" the nestlings and boil them down for a high quality oil that they used in cooking.

Oil birds nest in large colonies inside caves, and are the only fruit-eating birds to feed at night. They find their way in dark, confined places by sending out a series of clicking sounds and monitoring the echoes, on the same principle as the sonar of bats except that these sounds are within the range of human hearing. One evening when I was exploring a narrow rocky defile, the almost cavern-like silence of the forest was broken by several loud, sharp clicks. The dark shape of a broad-winged oil bird came close overhead, then another, and the clicking sound went echoing away as they disappeared along the canyon. I suspect that oil birds may be more numerous than is commonly supposed. I have often heard them at dusk in appropriate places, but they are not at all easy to observe except at their caves, and since these are hard to find, a belief in their scarcity has grown up.

When I arrived at Popayán, a fine Spanish town of white fronted houses standing 5,000 feet above the river Cauca, Marion had everything ready for our expedition into the *páramo*. In the icy cold open places above the forests, the land is useless for agriculture and the country is as wild as when Humboldt, Bonpland and Caldas explored it. The day after my arrival, we left Popayán and followed the route up into the *páramo* which Humboldt's little party is known to have taken. It leads towards a small, green-clad volcano named Puracé, which they climbed, about 25 miles east of the town. The road gradually deteriorated into a rough, Andean track. Where the track cut into the hillside, the banks were bright with flowers: purple and white foxglove-like penstemons, deep blue lupins only a foot or two high, and low bushes, belonging like the lupins to the pea family, covered with yellow, broom-like flowers.

We passed a few small settlements, usually the homes of no more than one or two Colombian families. Beside one roughly-built stone house, a man was digging a cultivated patch. His crop was potatoes, the tubers small and mis-shapen, but easily recognizable. He pulled them carefully from the dark soil and after brushing them clean, laid them on an old and much worn blanket. We talked for a while. He spoke good Spanish although he was probably descended from the Indian tribes who lived in the highlands when the Spanish arrived. The potato has an honoured place in local botany, for it was introduced to the rest of the world from the Andes. It probably originated in the central Andes of Peru and Bolivia and was commonly eaten by the Incas when the Spanish discovered their kingdom in the 16th Century. The Incas had even found a way of making instant potato, and the Indians of the central Andes still make chuño and tunta, two similar forms of dehydrated potato, produced by squeezing the potatoes dry in the daytime and freezing them at night. The process may take a month, but the almost weightless product, which can look very unpotato-like, will keep for years.

Soon we had ascended to the sub-páramo level at about 9,000 feet where small, waist-high trees are abundant. The surface of the track was broken by large pools of water from a recent rainstorm. Heavy grey cloud hung low on the hills around. Somewhere not far away to our right was the 16,000-foot peak of Puracé, but the cloud was so thick that we could see less than a hundred yards. The track levelled slightly and seemed unlikely to reach blue sky, so we decided to take the best trail we could find and continue more sharply up the slope on foot.

We had to stop frequently to find a way through the stunted trees and bushes, heavy with moisture. A film of cold water was soon flushing down the front of my anorak. To reach more open ground, I pushed slowly between long frail bamboos, and then followed a small stream flowing between moss-covered boulders. The ground everywhere was soft and springy with water, and a few places were treacherously boggy. Every square inch was thickly clothed with plants. The small outcrops of rock were covered with grey lichens. Many of the plants were grasses of the Calamagrostis and Festuca genera, the second related to the familiar fescue lawn grasses. Here plants even grew on plants: many of the low bushes were festooned with an epiphytic growth of straggly lichens, their spongy masses spangled with tiny drops of moisture.

As we picked our way through some bushes, we found that we were at the edge of a deep gully so filled with plants that hardly any light penetrated it. We forced our way between the moss-covered stems down a steep slope into the green, fog-wreathed tangle. The vegetation met overhead and the ground was thick with soft, wet moss. A small stream ran along the bottom of the gully, and the undergrowth on the opposite slope seemed too thick to penetrate. Climbing back up the way we had come was going to be difficult. So we stopped, rested and looked at the curious trap we were in.

Between the rotting stems of bamboos and other plants there was a comparatively clear space, and the stream ran only a few feet below us. The tunnel-like aspect of this open way caused me to suspect that we

On a grassy slope of the páramo a
scattered group of giant Espeletias, six
or eight feet high, stand out in the
sunlight against the inky blackness of
an advancing storm. The genus
Espeletia, which includes some smaller
species, is widespread in the drenched,
northern Andean uplands. Its giant
representatives are much the largest
plants to survive in the harsh páramo;
some grow to heights of 20 feet.

had entered a path made by some large animal. I listened intently, although I mentally admitted that our crashing struggles would have scared away anything in the vicinity—not entirely to my regret, as I did not fancy meeting anything large in such a confined space. The animal most likely to have made the pathway was the rare hairy or mountain tapir, first recorded near Puracé in 1829. The tapir is the heaviest mammal in South America and in the lowland forests can weigh as much as 400 lb. The mountain tapir has the heavy build and soft, extensile, trunk-like snout of all tapirs, but it is distinguished by the coarse hairs on its hide. I stopped and listened; happily, none seemed to be coming our way.

We took nearly half an hour climbing out of the gully, and when we reached the top, we were covered with a sodden mass of decaying vegetation. After labouring uphill for another 500 feet, I saw a group of tall figures looming through the mist ahead. They were the giant Espeletias known as *frailejones* or tall friars, one of the strangest plants of the northern Andes. Like many plants of the *páramo*, the Espeletias belong to the daisy family. This is the largest family of flowering plants and includes sunflowers, thistles, ragworts and even chrysanthemums. It was hard to accept the *frailejones* as members of the same family. They are much the largest plants of the *páramo*, and some of the largest species reach heights of between ten and 20 feet.

I approached one of the Espeletias for a closer look. The trunk, or perhaps one should call it the stem, was clothed with soft, dead leaves hanging in a succession of fringes, one above the other. A large rosette of grey-green lanceolate leaves crowned the top of the plant, and when I reached up and pulled one away, the central vein of the leaf cracked and oozed drops of liquid. The leaf was covered with silky hairs flattened to its surface like the soft fur on a rabbit's ears. From the centre of of the plant's crown two flower-stems protruded nearly a foot, bearing pale yellow flowers each formed from dozens of florets, like the composite flower of a dandelion. Only a few of the *frailejones* had flowers, though, and some of the largest seemed dead. I probed between the muff of dry leaves hanging around the trunk-like stem of one plant, and found that the material underneath was black and soft. When I touched the stem of another plant it fell to pieces. The tissue was like a wet sponge.

By now we had emerged into the *páramo* proper, and the vegetation clung close to the ground. The air was distinctly cold and it was hard to remember that we were only a couple of hundred miles north of the Equator. But then we had climbed above 10,000 feet. The mist seemed to be clearing and we saw other groups of *frailejones* standing on the slopes. The sun imparted a brilliant luminescent glow to the mist, bringing out the colours in the green carpet of the *páramo*: tubular red flowers, crimson bell-like blooms and pearly grey lichens. For the first time we noticed the gentle humming of insects. A few white, fragile, daytime moths appeared, and then a loud droning attracted my attention. I traced the sound to some low heather-like plants, where I found a large bumble bee. There are about 18 species of bumble bee in the Andes, almost the only place outside the northern hemisphere where they occur. The bumble bee concentrated my attention on the plants. They were closely crowded, and the higher we climbed, the more I found that they grew in a distinctive pattern. I began to realize that one species after another possessed leaves growing in perfect rosettes. Once I had noticed this, I saw more and more plants conforming to the pattern. Some rosettes were low and flat on the ground, others stood up from the ground in rigid spikes. One plant, about eight inches high, had its small soft leaves placed on the stem like the steps of a spiral staircase.

By keeping close to the ground, the plants of the *páramo* protect themselves from the wind, and to some extent from the cold, since the temperature fluctuates least at ground level. The compactness of the plants minimizes their surface area, so reducing possible heat loss, and the rosette shape allows as many leaves as possible to catch the fleeting sunlight. Heat loss is also reduced by the fine hairs on the leaves of some plants, which trap a protective layer of warm air at the leaf surface.

By now we were soaked to the skin. My tough boots had turned to slimy leather and I would have to wring them dry at the end of the day. As we gazed around through the lingering mist, we felt we had a good idea of what the *páramo* was like, but there was still one more sight to see before we turned back. At last the mist cleared and in front of us the *páramo* stretched for several miles to the green-clad slopes of a low peak. From its position, we knew it was Puracé, the volcano that Humboldt, Bonpland and Caldas had climbed. The tide of vegetation washed up its quiet sides nearly to the top, smothering the entire mountain except for the summit, which was white. Only snow can quench the rampant green growth of the northern Andes.

The Naturalist's Eye

For the first 200 years of Spanish colonial rule, the economic potential of the Andes was jealously protected from the prying eyes of other European powers. Foreign travellers were rarely permitted to journey freely; and because Spain had few competent investigators of her own to send, little information about the Andes' plant and animal life came back to Europe. It was not until the late 18th Century, with its emphasis on reason and knowledge, that more explorers were allowed access to the Andean wild places.

The first scientific expedition commissioned to study the plant life arrived in Lima in 1778. It was set up and financed by the Spanish, in response to persistent requests from the French that they should be allowed to send a botanist to Peru. Thus the expedition's leader was an experienced French naturalist, Joseph Dombey, who was aided by two young Spaniards, Hipólito Ruíz and José Pavón. The three spent ten years journeying into the mountain forests, supported by Indian bearers and mule trains. They pressed and had drawings made of many thousands of plant specimens, which they sent back to Europe. They also produced accounts of the plants'

properties: how they healed, cased childbirth, befuddled the senses.

The exploits of one man, Baron Alexander von Humboldt, became most famous. Not content with having explored 1,700 miles of the river Orinoco, Humboldt set out in 1801 with his companion, Aimé Bonpland, to travel along the Andes from Colombia to Peru. He climbed almost to the summit of Chimborazo—at over 20,000 feet the highest mountain then known—and recorded in detail the vegetation on its slopes. Humboldt visited Andean regions previously only poorly investigated. He described and illustrated many new plant species and corrected the zoological and botanical errors of previous explorers. Indeed, Humboldt collected so much information that its publication took 20 years.

In the early 19th Century, a new dimension to the knowledge of Andean life was added by the French naturalist and palaeontologist, Alcide Dessalines D'Orbigny, who was the first to use fossils as a guide to the past. D'Orbigny, though, was at home in all branches of zoology and botany. His versatility, and that of many other early explorers of the Andes, provided some of the graphic evidence reproduced on these pages.

Humboldt portrayed this male Andean condor (right) in his "Voyage of Humboldt and Bonpland". He was amazed to see these fierce-looking birds gliding thousands of feet above sea level. Humboldt reported, however, that the condor's ferocity had been exaggerated by others. It neither carried off babies nor was it a danger to men, and it "prefers carrion to live animals".

This "lion-monkey", no larger than an ear of maize, darted about its cage so rapidly that Humboldt had difficulty drawing it.

Conqueror of the Andes

Alexander von Humboldt's conquests over nature were compared in his own lifetime to the military conquests of his contemporary, Napoleon Bonaparte. By exploring widely and observing every kind of natural phenomenon, Humboldt hoped to reveal an inner unity running through all parts of the physical world. The monumental work in which he attempted to set out the pattern was the *Cosmos*. The first volume was published in 1845 when he was 76. The fifth and last volume was only half completed when he died at the age of 90, and was finished by his followers.

The attention to detail that characterized all of Humboldt's work is seen in his illustration of the Andean condor on the previous page. Apart from giving the bird's length (three feet, three inches) and wing-span (nine feet, four inches), he specified even the lengths of toes—with and without talons.

Humboldt also paid close attention to the behaviour of animals, including the two monkeys shown on these pages. Of the howler, he wrote that it was placid in captivity, while the seven-inch-tall "lion-monkey" or pigmy marmoset was, "like most small creatures", irritable. When it was angry its mane bristled, making it look even more lion-like.

Humboldt counted 40 howler monkeys like this in a single tree. He reported that he often heard their howls as far as three miles away.

Cinchona grandiflora's bark, when ground up, was used to treat malaria and dysentery.

Botanists to the King

The Spanish botanical expedition sent by Carlos III to Peru in the 1770s and 1780s, under the leadership of Hipólito Ruíz, was lavishly equipped and precisely instructed. Its purpose was "not only to promote the progress of the physical sciences, but also to banish doubts and adulterations that are found in medicine".

One result of the expedition was the large work *Quinologia*, which described the varieties of the vitally important Cinchona tree. This tree provided quinine, which was the only remedy against malaria.

The uses of many other Andean plants were detailed in the expedition's *Florae Peruvianae et Chilensis* of 1794, from which the illustrations on these pages were taken.

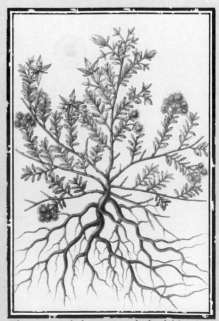

The roots of the ratana halted bleeding.

The guava tree produced fruits that the Andean Indians made into a "very good candy resembling pear jelly".

An Enchanted Collector

The journals of Alcide Dessalines D'Orbigny, the brilliant young naturalist despatched to South America by the Paris *Académie des Sciences* in 1826, breathe an air of pure delight into Andean natural history. They not only record the fossils collected by the author, but also demonstrate his fascination with all plant, animal and human life.

D'Orbigny was most intrigued by the birds that fluttered tantalizingly out of reach. At times he pursued the screeching toucans and numerous Cacique birds, hoping to shoot down specimens; "but in every case I was obliged to wait until these birds would descend to the second level of vegetation, as my firearms could not reach the treetops". Exploring the dark vaults of greenery, where brilliant flocks of insects fluttered around the flowers of palm trees, D'Orbigny was enchanted. "I don't believe," he wrote, "I was ever so happy in any place."

The kinship of the white-fronted woodpecker (top) and the flicker is detailed by D'Orbigny.

D'Orbigny had to scramble up sheer crags to find the mountain caracara (top) and its young. The eggs are those of a chimango caracara.

3/ The Uneasy Cordilleras

There is something surprising in the tranquillity of this deserted landscape where once a thousand volcanoes boomed to each other in their great subterranean organs and spat forth their fire. ANTOINE DE SAINT-EXUPERY/ WIND, SAND AND STARS

For almost 2,000 miles, from the border between Ecuador and Peru down to central Chile, the Pacific coast of the Andes is dry. As you travel south along the narrow strip at the foot of the Western Cordillera, the barrenness grows more and more extreme until, in the Atacama Desert of Chile, you reach what is probably the driest place in the world. There are parts of this desert where no rain has ever been recorded. The ground is totally sterile and devoid even of the hardy scrub found in other deserts, including the Sahara. In the Atacama's major coastal towns, such as Antofagasta, heavy rain falls only two or three times a century and water has to be piped from the snowfields.

Everything is preserved in the Atacama as if in dry storage. Often I have rediscovered my campsites of a few years ago exactly as I had left them. I have walked into mining camps that were abandoned nearly a hundred years ago but are still strewn with unrusted tin cans. On one journey I uncovered the relics of hunters who had lived in the desert thousands of years ago. There were pieces of clothing, hand-made ropes and mummified human remains, shrunken and leathery as parchment.

Behind this barren coastline, the great Western Cordillera rises to heights of over 20,000 feet, and among its peaks are some of the world's highest volcanoes. Many are active, and the chief purpose of my journey into this region of the Andes was to try to scale Ubinas, the most active volcano in Peru.

The great size of this brooding range partly explains the dryness of the Pacific coast, for it forms a last great barrier to the moisture-laden winds that blow from the Amazon basin. All their moisture is shed on the eastern slopes of the Andes, leaving a "rain shadow" on the western side. Even more a cause of drought, though, is the cold Humboldt Current which sweeps up the coast from the Antarctic. Alexander von Humboldt discovered it and its effects when he visited the port of Callao, near Lima, in the winter of 1803.

He was puzzled by the coastal climate—mostly grey, cool and foggy, yet virtually rainless. When he took temperature readings, he found something most unusual about the sea breeze. Normally, wind blowing from the sea is warm and moist. It cools over the land, condenses, then releases rain. In the coastal desert, by contrast, the wind from the sea is cool, having passed over the waters of the cold Humboldt Current. As it blows over the coastal strip, therefore, it is warmed by the land; and because warm air can hold more moisture than cold, it actually sucks moisture from the land. This explains the extraordinary paradox of a desert shrouded in fog. Only in a very few places is the fog heavy enough to precipitate rain or support life. Without precipitation, the ground remains parched.

On my way to Ubinas, I travelled through the desert on the narrow Pan-American highway which follows the coast, often winding up along cliffs hundreds of feet above the sea. Near the small town of Camaná I stopped to explore a rocky headland where the stony, barren foothills of the Andes plunge abruptly into the Pacific. At once I noticed the pronounced cold breeze blowing in from the sea. But my attention was quickly diverted by the astonishing proliferation of seabirds on the rocky cliffs. Cormorants, their white fronts like dress shirts, stood shoulder to shoulder, jostling each other for room. Innumerable pelicans and elegant boobies with mottled, grey-brown wings competed with the cormorants, presenting a scene of frenzied activity. Streams of parent birds flew off to feed, so many that it seemed the cliffs would soon be bare. Yet immediately, their places were taken by others landing with food for their young. All along the coast, similar sites have been used by seabirds for centuries, and in some places the droppings formed deposits of guano that accumulated to depths of hundreds of feet until they became world famous and were exploited for fertilizer from the 19th Century onwards.

The millions of seabirds are one of the animal wonders of the world, and they provide a clue to another remarkable feature of this coastline:

the Peru-Chile Trench. They feed on the vast shoals of anchovy which are found in the cold, grey-blue waters of the Humboldt Current. The anchovy in turn feed on plankton, and the plankton flourishes on an exceptionally rich concentration of dissolved phosphates. These minerals are drawn up in giant eddies from the chilled, salty depths close to the coast, where the seabed plunges sharply to a depth of some five miles—farther below sea level than the highest Andean peaks are above it. In places the distance between mountain top and sea floor is nine miles. This is one of the longest ocean trenches in the world. It stretches from close to the Equator right down to southern Chile, closely hugging the line of the shore. West of Antofagasta in Chile it reaches a depth of 25,000 feet within 50 miles of the coast.

Before continuing along the Pan-American highway, I spent some hours exploring the headland near Camaná—200 miles from Ubinas—watching the ceaseless comings and goings of the birds. The powerful oceanic swell sent up plumes of spray that drenched the rocky shore and periodically blew a fine, salty mist into my face. It was a dramatic scene, and for me it was made even more powerful by the knowledge that I was standing at the brink of one of the most unstable sections on the earth's crust.

It is no coincidence that the Peru-Chile Trench lies so close to the Andean shore. The depth of the sea and the height of the mountains are intimately related. Indeed, the trench is the birthmark of the Andes. The tectonic movements by which it was formed have also created the mountains, the great volcanoes and the spasmodic Andean earthquakes.

All these phenomena are signs of a collision between two gigantic sections of the earth's crust. It is surely one of the most exciting discoveries of the 20th Century that the outer rind of the earth, the lithosphere, consists of separate pieces, like a globe-shaped jig-saw puzzle, and that those pieces move. According to the revolutionary theory of plate tectonics, there are six major pieces, or plates, and many smaller ones which move around on a layer a hundred miles down in the earth's mantle. One of these plates comprises, roughly speaking, the floor of the east Pacific Ocean. Next to it is another large plate that extends from the Pacific coast of South America to the mid-Atlantic, carrying the entire continent as a passenger on its back.

No one knows exactly how the plates move, but the impetus seems to come from the oceanic ridges that lie between them. If the Atlantic Ocean were to be drained, a great mountain range would be revealed running right down its centre, studded with active volcanoes. Magma

Surrounded by a horde of cormorants on the Peruvian coast, a red-headed turkey vulture tears at the flesh of a dying bird it has found.

from the earth's mantle is continually injected into this ridge, attaching itself to the plates on either side. As the plates are enlarged in this way by the solidifying magma, they move apart and travel away from the mid-oceanic ridge like conveyor belts inching in opposite directions.

In this way South America has been carried steadily westwards on its plate during the last 200 million years. At the start of this period— very recently in the 4,600 million years of the earth's history—it was still joined to Africa as part of the ancient super-continent of Gondwanaland which also embraced India and Australia, and its west coast was flat. Since then South America has moved 3,000 miles away from Africa and it is still inching westwards. Meanwhile, the floor of the Pacific has been travelling eastwards towards it, and the result has been a gigantic collision between the two plates.

As the east-moving Pacific, or Nazca, plate (named after a valley in the Andean coastal desert) met the west-moving South American plate, something had to give. The Nazca plate was forced down into the earth's hot mantle, fusing itself into magma once more and leaving a giant abyss, the Peru-Chile Trench. Over 200 million years this mighty geological collision caused the flat coastline of South America to buckle and fold. Magma meanwhile welled up to the surface, erupting through the vents of a thousand volcanoes. Gradually a range of young mountains was formed: the Andes.

Visible evidence of this continuing geological process can be seen in the volcanoes that lie dormant or fume ominously along the length of the range, from Colombia to the southern tip of the continent. Most of the volcanoes—and the most active—stand in the Western Cordillera behind the long desert coast. In Chile alone, 22 are regarded as active. In southern Bolivia and northern Chile there exists probably the greatest density of volcanoes in the world. In Ecuador, where the Andes narrow to less than 60 miles wide in places, I have seen ten major volcanoes in a day. One of them, Chimborazo, rises to 20,577 feet. Before the Himalayas were surveyed it was thought to be the world's highest mountain. Not far away, Sangay (17,450 feet) is one of the world's most active volcanoes, lighting up the clouds at night with a lurid red glow.

Humboldt climbed many of Ecuador's volcanoes before heading down to Peru to make his observations about the cold current and its effect on the coastal climate. He found that Pichincha, which overlooks Quito, the capital of Ecuador, was active and excitedly recorded the rumblings and explosions in the summit crater. When he returned to Quito the suspicious inhabitants accused him of causing the explosions

Halfway along the Andes, at the junction of Peru, Bolivia and Chile, the mountains divide into the Western Cordillera and the high inland cordilleras of the east (brown) with peaks of 16,000 feet or more (white). Cupped in this division is the Altiplano, a cold and bleak basin at 12,500 feet which once held two vast inland seas. All that remains of these ancient bodies of water are salt flats and Lakes Titicaca and Poopó. The outer slopes of the cordilleras (buff) descend to their foothills at about 1,500 feet, where they gradually flatten out into the coastal and jungle lowlands (green).

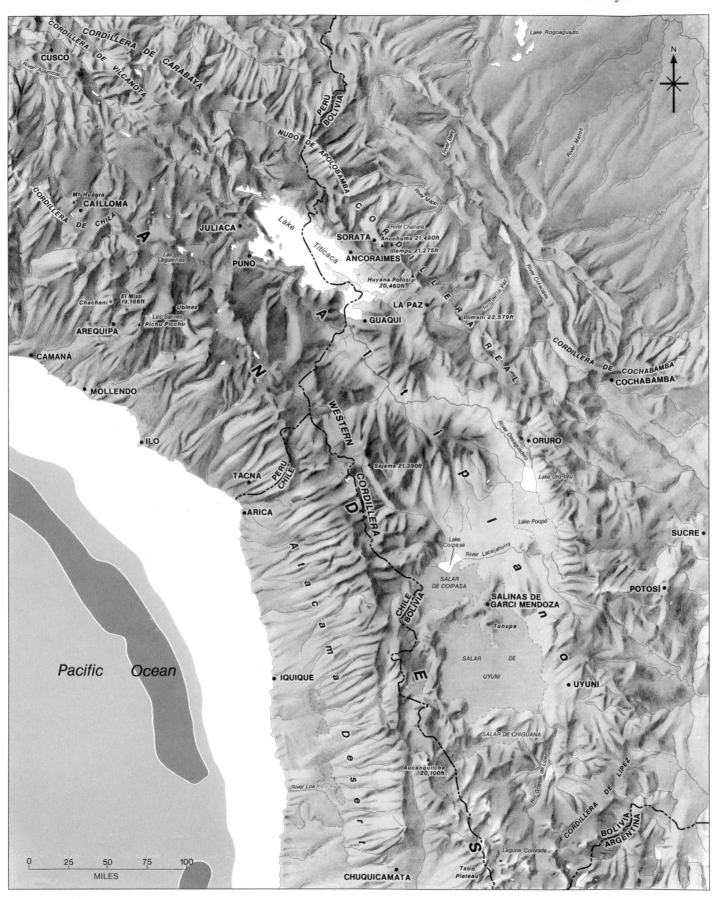

CORDILLERA DE VILCANOTA
CORDILLERA DE CARABAYA
CUSCO
River Apurimac
Lake Rogoaguado
N

NUDO DE APOLOBAMBA
PERU
BOLIVIA
River Beni

River Matos

River Mapiri

Mt. Huagra
CAILLOMA
CORDILLERA DE CHILA
JULIACA
River Challana
SORATA
Ancohuma 21,490ft
Illampu 21,275ft
ANCORAIMES
C
O
R
D
I
L
L
E
R
A

Lake
Titicaca
PUNO
Las Lagunillas
Huyana Potosi 20,460ft.
El Misti 19,166ft.
Chachani
Ubinas
LA PAZ
Rio de la Paz
River Cotacajes
CORDILLERA DE COCHABAMBA
COCHABAMBA
Las Salinas
Pichu Picchu
GUAQUI
Illimani 22,579ft
R
E
A
L

AREQUIPA
A
N
D

CAMANÁ

MOLLENDO
l
t
i

ILO
River Desaguadero
ORURO

Sajama 21,390ft
P
Lake Uru-Uru

TACNA
PERU
CHILE
WESTERN
SUCRE
ARICA
l
a
Lake Poopó
Lake Coipasa
River Lacajahuira

CORDILLERA
POTOSÍ

SALAR DE COIPASA
SALINAS DE GARCI MENDOZA
n
o

IQUIQUE
Atacama
CHILE
BOLIVIA
Tunupa
SALAR DE UYUNI
UYUNI

Pacific   Ocean

D
E
S
Desert
SALAR DE CHIGUANA

Aucanquilcha 20,100ft
Rio Grande de Lipez

River Loa

CORDILLERA DE LIPEZ
BOLIVIA
ARGENTINA

Laguna Colorada
Tatio Plateau

CHUQUICAMATA

0   25   50   75   100
MILES

himself by throwing quantities of gunpowder into Pichincha's crater.

So great is the geological activity in the western Andes that it is almost impossible to spend a month there without experiencing an earthquake. For someone like myself, not brought up in this unstable part of the world, the first tremor leaves a unique and lasting impression. It is as though the earth on which you have relied since birth can no longer be trusted. Just before the earthquake there is an unnatural silence, a presage of disaster. I experienced my first tremor about 200 miles north of Camaná, where I was staying in a tiny Indian village on the western slopes. The silence came during the middle of supper. The Indian woman in the corner of the room suddenly stopped jogging the baby on her back. Almost in a whisper, she said: "Temblor, señor; un temblor." A little ripple appeared in my soup. Instantly the building shook. First there was a threatening rumble deep in the earth, then an all-encompassing roar. In seconds it stopped. Then we were outside, away from the house. We waited several minutes, expecting more, but the ground had settled and become solid again.

Earthquakes are not isolated movements but wave motions that can sometimes be heard rushing towards you and then rumbling past like an express train. Thousands of tremors are so small that they can be detected only by a seismometer, but the big ones can cause devastation on a vast scale. In the Peruvian town of Yungay, 20,000 people died when an earthquake started an enormous landslide in 1970. Yungay was obliterated by a massive wall of ice mixed with boulders and mud. The only survivors were those who happened to have been in the little cemetery that stands on the hill above the town.

A cliff of volcanic ash displays bands of colour representing successive eruptions of El Misti and neighbouring volcanoes. The ash deposits consist mainly of small lumps of pumice—a light-weight, white stone shot through with bubble holes. The darker coloured strata are also pumice, but they have been sullied by soil and mud.

The last thing anyone wants to experience is a major earthquake, but an almost equally dramatic aspect of mountain building is an active volcano. I had always looked for an excuse to climb one and see what was happening inside the crater. The chance came when I was on my way through the coastal desert. There were reports in the local newspapers that Ubinas, which stands inland from Camaná, was distinctly active. One group of mountaineers had said that the volcano was making tremendous noises and that "a green lake of boiling water" filled its crater. I determined to see it for myself.

After leaving Camaná, I drove up into the mountains. At first the countryside was shrouded in fog, but at about 2,500 feet I emerged suddenly into clear air and could look back over the top of the fog as if I were looking down on a layer of cloud from an aeroplane. Ahead was a shimmering desert leading to the distant mountains, to the city of Arequipa 8,000 feet up, and to Ubinas.

In Arequipa I joined Marion, who had arranged a rendezvous with Dr. Peter Francis, a British geologist who had specialized for many years in Andean volcanoes and was to accompany us on the trip. After spending two days in Arequipa in order to acclimatize ourselves to the high altitude, we set off by Land Rover into the volcanic cordilleras. We began by driving along a dusty tortuous track. Slowly, climbing all the time, we skirted the weathered form of Nevado Picchu Picchu, known as the sleeping Indian from its profile against the sky. When we reached 12,000 feet, we could see the summit of the volcano El Misti about five miles away to our left. The sky was clear and every detail of its 19,000-foot, classically shaped cone was visible.

Our road cut through deposits of ash that had been thrown out by the volcano at some time in the past 50,000 years. The different kinds of material ejected by the explosions showed clearly as multi-coloured layers, some of them up to 20 feet thick. We stopped the Land Rover and scrambled uphill to one of the more dramatic layers. I picked up a small piece of light, pinkish rock. Its surface was jagged and harsh looking, but I could score it with my thumbnail. "That's pumice," said Peter. "The honeycomb effect was caused by hot gases escaping from the lava."

Most of the material around us was pumice and ash. This was to be expected, for Andean volcanoes emit relatively few lava flows of the familiar type. Peter explained a typical Andean eruption. "Imagine a rather warm, well shaken bottle of beer," he said. "When you take the top off, a jet of liquid foam spurts high into the air and then falls back: that's ash, which is what we see here. Some frothy liquid wells up and runs down the side of the bottle: that's ignimbrite." Ignimbrite is a more solid material that hardens into a foamy, white rock. We could not see any around us now, but Peter thought there would be some from earlier eruptions buried beneath the ash. "The stuff that stays in the bottle," continued Peter, "is andesite, a type of lava somewhere between basalt and granite, and too viscous to flow freely. Just occasionally it comes up too, in the form of very sluggish lava flows."

The energy of the explosive eruption that had showered the ash around us was clearly stupendous—greater, Peter said, than any hydrogen bomb. But this eruption was like a damp squib compared with those which began in this area about 15 million years ago. At that time there was a series of violent eruptions, each larger than the explosion of the Indonesian volcano, Krakatoa, in 1883. Immense quantities of ash were sprayed into the air, accompanied by huge flows of ignimbrite. These ignimbrite flows were deadly. They surged up from the volcanic vents, a froth of ash particles buoyed up by incandescent gases, and flooded out over the countryside at speeds of up to 60 miles an hour, vapourizing every living thing in their path. The period of ignimbrite eruptions seems to have lasted until about two million years ago, and the resulting sheets of ignimbrite cover nearly 70,000 square miles of the central Andes to thicknesses ranging from inches to more than 1,000 feet.

The rocks deposited by the ignimbrite flows arc known locally as *sillar*. The city of Arequipa was built (and rebuilt after earthquakes devastated it in 1958 and 1960) almost entirely of *sillar* which erupted from one of El Misti's ancestors. The rock is light yet strong; soft enough to be sawn with simple tools, yet durable enough to last for centuries. The craftsmen of Arequipa take advantage of its softness to carve it into all manner of forms. Even the façades of banks and cinemas are covered with intricate coats of arms, floral symbols and geometrical shapes, all made of *sillar* and standing out brilliantly in the hard Andean sunlight.

We drove on across a bleak, semi-desert plateau, stopping to ask the way at an isolated Indian house thatched with tough mountain grasses. "Señores," one of the Indians asked, "what is happening there at Ubinas? Are the mountain spirits angry?" He was anxious about his crops and his animals, for a great ash fall more than three centuries ago had laid waste the entire area around us. This was probably the major eruption in 1600 recorded by the Spanish traveller, Vásquez de Espinoza. According to Espinoza, Ubinas showered ash for more than 200 miles around, some of it falling on ships at sea, and there was so much red-hot pumice in the rivers that the fish were boiled.

The Indian directed us towards a track that led upwards to our right, due east of El Misti. He said we would be able to see Ubinas on the skyline as we passed Las Salinas, a salt lake also to our right. Marion took the wheel and we set off again across the plateau. The Indian was right. Soon after the gleaming white surface of Las Salinas came into view we caught sight of the distinctive profile of Ubinas; not a simple, elegant cone like El Misti but rather irregular with a broad, flat top. Although in shape Ubinas does not look like everyone's idea of a volcano, there could be no mistaking the faint, white wisps of fumarolic vapour rising above it. After a long drive across the plateau, with Ubinas always looming ahead of us, we finally arrived near its foot just before dark. The altimeter showed 14,000 feet.

As the sun set, a range of volcanic hills in the west cast a shadow on to

In the mid-morning sun, heavy shadows are cast by andesite boulders strewn on the desolate, western slopes of the volcano Ubinas, 15,000 feet above sea level. Broken off from the main lava flows by freeze-thaw erosion, the boulders have slid down the steep gradient over thousands of years.

the side of Ubinas. We watched the volcano's slopes slowly darken and felt the chill of the mountain night. It was hard to believe, as Peter and I hurriedly made camp in the icy dusk, that a few hundred feet above us, where Ubinas was bathed in a golden glow, the day was still warm. Already the water in our jerry cans was freezing. Within moments we had gathered enough scrub to get a fire going. We had soup, and then quickly got into our sleeping bags.

It was 6 a.m. when we braced ourselves to meet the cold. At once we were reminded of the gaseous activity in the volcano's crater above us. Acrid, sulphurous air wafted down, stinging our nostrils and leaving an unpleasant taste in our mouths. The wind must have changed, for we had not noticed any smell the night before. The sky was just getting light, but the rising sun was still behind Ubinas. We hastily lit another fire to brew coffee. Everything was frozen solid, and it was not until the fire had been burning for several minutes that we were able to thaw and empty the cups of the dregs from the night before. We set out at 6.45, still in the icy shadow of Ubinas, and walked briskly to get warm. For me, the blanched, dusty sides of the volcano offered something of a challenge. I had broken both my legs in a serious accident in the Andes two years before. I could now cope with level ground, but the mountain that towered above us was a new and totally unknown quantity.

At the start, the gradient was gentle and the ground firm and easy to walk on. Twice we had to climb down into steep gullies and out again before continuing upwards, yet in the first hour we ascended almost a thousand feet. Still the sun had not reached us and ice and snow lay unmelted between the rocks.

The next hour took us over steeper ground. In one place we had to climb a slope which was so steep and sandy that to gain a foot meant scrambling perhaps two or three. Here the altitude began to trouble us for the first time. Two days of acclimatization in Arequipa had not been enough, and the rarefied air on the slopes of Ubinas were sapping our strength. Fine, white dust from previous eruptions covered the volcano's side, so that it resembled an interminable sand dune. The slippery slope was interrupted only by piles of burnished rock.

Finally the sun rose above Ubinas. The intense cold of the night and early morning was now banished by an equally intense heat, which was already beginning to burn our faces in spite of the thick layers of high-altitude cream we had applied before setting off. The static electricity in this thin, dry air made the white dust cling to everything and the anti-static brushes for the camera lenses were useless.

We pushed on up the ever-steepening gradient, breathing hard now. Around us we noticed the fractured remains of many black, andesite "bombs" left over, probably, from a prehistoric eruption. They had been hurled out of the volcano as blobs of molten lava. Then they had hardened in the air and crashed to the ground. We also found pancake-like circles of lighter rock—more fluid lumps that had flattened on impact. Amid all these signs of some hellish bombardment, we were thankful that the crater above us was only giving off vapours.

We stopped for rest, well above 16,000 feet now and breathing heavily. The air was even thinner, the slope had increased unpleasantly, and the muscles in my legs were tiring. We went on again. After every 20 paces or so we were gasping for breath and had to rest. The urge to give up was strong, but the crater was only 300 feet above us, and the prospect of the view from its rim was too tantalizing to resist.

There was now a choice of routes before us. Either we could make a frontal assault, continuing up the slope, or we could traverse a boulder-strewn area to our left and climb up a gully strewn with large chunks of shattered lava. As it turned out, we made the wrong decision. We pushed straight up towards the crater's rim. The surface was so soft that each step gained us less than six inches, sometimes nothing. And sometimes a step in the wrong place meant a long slide downhill.

Marion, who was well ahead, yelled down to say that the slope showed

no sign of improving. Although she was still going strongly, we reluct-
antly decided that we would all turn back. Peter's altimeter showed
nearly 17,000 feet. I sat down, or more truly sagged, against a convenient
rock and tried to regain my breath. With more time we could all have
coped with the thin air, but clearly the only prudent choice was to
descend since my speed was limited and we did not relish the thought of
a night camped on the summit.

There are two good things about resting. It is, of course, pleasant; but
it also gives you a chance to look around. Now for the first time I could
take in the amazing scene of desolation below me. The side of Ubinas
was white and blinding in the hard sunlight, forcing me partly to close
my eyes. The land was dry and barren. Every fold of the Andes was exposed, its surface deeply incised and weathered. I could see the white
expanse of Las Salinas, and in the distance the neat cone of El Misti. Far
below, the rocky chaos reached down to sandy desert; and beyond that,
lost partly in the haze, was the Pacific.

Although the sun was higher now, the air was pleasantly cool at this
altitude. To the east I could see a line of magnificent, snow-capped
mountains: the older, eastern cordilleras of the Andes. All around me,
the land was silent and still, save for a pale yellow butterfly that passed a
few feet in front of me. For an instant the landscape seemed remote, dis-
connected from my presence. Time seemed to stop, and I felt I could see
all the ages of the Andes simultaneously.

A waft of sulphurous air drifted down from the crater. Ubinas had
beaten us, but the view from its summit was a fine consolation, a panorama of the young Andes. As we began the trek downhill to our camp,
I could imagine this stretch of land 200 million years ago as a flat coastline. I thought of the Nazca plate moving inexorably down into the
earth's mantle beneath us. It had created this contorted landscape and
was still working its changes. In some months, from far north to far
south, the Andes betray no activity. In others there may be a volcano
belching smoke, fire and ash in Chile or an earthquake somewhere in
Peru or Colombia. For the moment, here, all was quiet.

The Seething Peaks

PHOTOGRAPHS BY LOREN McINTYRE

From Colombia to Cape Horn, more than 40 active volcanoes, some soaring to over 20,000 feet, testify to the deep subterranean forces still shaping the Andes. They are grouped in three principal regions, in the green mountains of southern Colombia and Ecuador, the deserts of southern Peru and northern Chile, and the icy extremities of southernmost Chile. These active regions are linked by other, extinct cones to form the longest and highest chain on the active volcanic belt, or "girdle of fire", that encircles the Pacific Ocean.

One of the most active groups lies along the twin cordilleras of Ecuador. Here, only a few miles south of the Equator, Cotopaxi's symmetrical cone (opposite) rises to such a height that glaciers shroud its upper slopes, adding a deadly hazard to the volcano's eruptions. In 1877, during an unusually severe upheaval, volcanic materials cascading on to the ice produced a flash flood, and the resulting mud flow inundated valleys as much as 200 miles away.

The glaciers were probably melted by heavy falls of hot ash, ejected during a typically Ecuadorean eruption. The same kind of eruption can be seen nowadays about 100 miles south of Cotopaxi. Here the most active volcano in the Andes, Sangay, spews up shrapnel bursts of lava, ash and steam almost daily. But from time to time, with a versatility typical of Ecuadorean volcanoes, it emits molten lava which wells up in its summit crater and spills down its grey, clinker-strewn flanks.

Farther south, a series of giant but less active volcanoes towers over the eastern slopes of the Atacama Desert, exhaling sulphurous fumes into the parched air. They have no glaciers on their summits and little rain falls to wash away the ash and pumice beds that mantle the volcanic slopes. As a result, on some cones ancient lava flows are preserved intact beneath the younger deposits, providing vulcanologists with a rare opportunity to trace a complete record of past eruptions.

South of Aconcagua, the highest peak in the Andes, the volcanoes of Chile match those of Ecuador in size and activity, dominating a maze of knife-edged ridges and plunging valleys. Then, as they approach Cape Horn, they become smaller and their jagged peaks emerge from thick ice caps. In this bleak and little explored region, the landscape has been reduced to its most stark and primitive elements: ice and fire.

An aerial view of Cotopaxi's snow-covered summit crater captures the lethal symmetry of this slumbering monster, which last staged a major eruption in 1942. The symmetrical form has been built up by the ejection of volcanic material from a single, centrally placed vent, clearly visible in the crater.

Sangay, the so-called "flaming terror" of the Ecuadorean Andes, belches a cloud of sulphur-laden gases, relieving the pressures within.

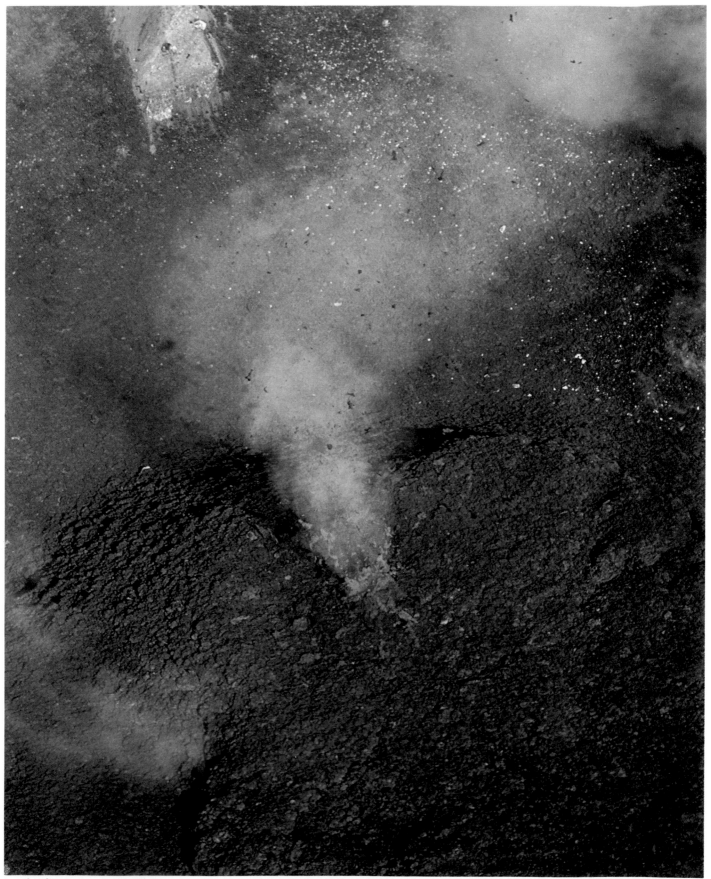

A subsidiary vent on Sangay's crater lip ejects molten lava which, falling back, fuses in an irregular mass to form a low spatter cone.

Tocorpuri guards the eastern rim of the Atacama Desert, looking out over a desolate landscape of extinct volcanoes and salt-flats. On its scarred and pitted 18,000-foot summit, a fumarole breathes a feeble column of gases into the rarefied air. Sulphur beds, formed by the chemical reduction of the gases, lie in yellow patches, surviving in the arid climate as valuable mineral deposits.

In a remote part of southern Chile, two dome-shaped cones close off a small valley, one of them forming a dam for an icy mountain lake. The cones are composed of loose cinders, or scoria, which were ejected in the first stages of eruption. Shortly afterwards, lava flowed from the base of the cones and flooded down the valley, piling up into gigantic knots and ripples as it cooled.

4/ A Rarefied Existence

Not all the cordilleras behind the desert coast contain volcanoes. Some
are quiet, granite mountains, among them the great Cordillera Blanca in
Central Peru, a spectacularly high, snow-capped range with 22 peaks
of more than 19,500 feet, one of which, Huascarán (22,205 feet) is Peru's
highest. I was drawn to the Cordillera Blanca because it is one of the
best places to explore the cold, grassy steppe-land or *puna* which
clothes the high Andean slopes between 12,000 feet and the snow-line
at 16,500 feet. The *puna* is a harsh environment, swept by high winds,
without rain for nine months of the year and bitterly cold at night.
During the day the sun is warm yet the ground in the shadows remains
frozen. In the thin, stony soil, little grows except clumps of coarse,
yellowing grasses and dwarf plants hugging the ground. There are
virtually no trees and the only tall plants are the slow-growing, cactus-
like puyas, perhaps the strangest relatives of the pineapple and confined to a few parts of the high Andes.

Not surprisingly in this hostile environment, animals are rare. Those
that do survive possess a variety of physiological and behavioural
adaptations. They tend to reproduce slowly, suiting their populations
to the scant resources of the *puna*. Many are rodents, lizards and birds
small enough to find cover in the low vegetation, among rocks or in
burrows and tunnels underground. Their food needs are slight, and
although small bodies lose heat faster than large, their ability to find

shelter protects them from the cold. Some species of beetles and birds are known to huddle together at night in self-insulating communities.

There are few large animals in the *puna*. Apart from the puma, or mountain lion, and an antlered deer, the *taruca*, both rare, the only good-sized mammals are members of the camel family, the guanaco and the fast-running vicuña, both wild, and the domesticated llama and alpaca. Although they cannot find shelter easily, they are protected from the elements in other ways. With their larger body mass, they lose heat less rapidly than the smaller animals. They have a covering of fine, fluffy wool, for which the vicuña in particular is famous, and this traps a protective layer of air, providing excellent insulation. Even these animals, though, are small in their own family, weighing barely a third as much as a Bactrian camel. The giant in the *puna* is the Andean condor, the largest bird of prey in the world and found only in these mountains. Its great wingspan, which has been known to reach ten feet, enables it to soar effortlessly to 18,000 feet on upcurrents of air, covering large distances with little exertion and so finding sufficient food.

One of the biggest problems above 12,500 feet is the altitude itself. The sunlight at these heights is intense. Most insects are coloured in dark browns, reds, blacks or deeper-than-average yellows—dense pigments which afford protection against ultra-violet rays and other solar radiation, and also absorb heat more readily than lighter shades. In the high mountains, atmospheric pressure falls progressively until, at 18,000 feet, the oxygen in the air is roughly half the amount at sea level. The vicuña, which relies on its speed for escape and for long-range foraging, is adapted to the thin air by having an enlarged heart and lungs and an exceptionally high level of oxygen-carrying red corpuscles in the blood—about eight million per cubic millimetre. The Quechua Indians, who have lived at heights in excess of 12,000 feet for centuries, are adapted in the same way. Their barrel-shaped chests enclose hearts that are enlarged to circulate an extra 20 per cent of blood.

For anyone coming from lower altitudes, the initial effects of the high mountain atmosphere can be crippling: headaches, quick breathing, a thumping heart, fainting and nausea are all symptoms of a condition known in Peru as *soroche*. Most people, depending on their physical fitness, can acclimatize themselves in a few days, during which time the body builds up more red corpuscles; some do not become acclimatized at all, and can die of embolism. During our ascent of Ubinas we had seen how two days of acclimatization at 8,000 feet had not been enough to prepare us for strenuous exertion at high altitude.

In the Cordillera Blanca, Marion and I would be above 12,000 feet for a whole week, and I wondered how we would fare. We intended to climb into the *puna* from the narrow, beautiful valley of the river Santa, the Callejón de Huaylas—*callejón* being the Spanish for corridor, while Huaylas is a small town at the northern end of the valley, where the river breaks from the mountains through a canyon 30 feet wide. The luxuriance of the lower slopes, cultivated for many generations by the Quechua, contrasts sharply with the harsh *puna*. Fields of ripening maize cling to the mountainside like yellow postage stamps, and 12,000 feet above the floor of the valley—which is itself 10,000 feet high—tower the mountains, the Cordillera Negra to the west, so called because its peaks are often snowless, and the white-capped Cordillera Blanca, our ultimate destination, on the east.

To reach the Callejón de Huaylas, Marion and I drove inland from Lima. I was uncertain how the valley would look, for in the Peruvian earthquake of 1970, just a few years before, it had been the main disaster area. Yungay, the town which was destroyed with the loss of 20,000 lives, lay on our route 30 miles south of Huaylas, at the foot of Huascarán. All the villages we passed through showed signs of the earthquake. Sometimes only the walls of houses were cracked or leaning at odd angles, sometimes entire districts had been reduced to rubble. When we reached Yungay, we found the town—or what was left of it— buried by mud. All that showed of Yungay were the tops of the palm trees in the main *plaza* and part of the cathedral tower.

Travelling south down the Callejón, we reached the village of Catac where we found a stocky Quechua guide who agreed to take us into the *puna* and promised to show us some giant puya plants in flower. *Puya raimondii*, named after Antonio Raimondi, the Italian naturalist who discovered it in the mid-19th Century, is thought to be a relict plant from the time when armadillo-like glyptodons and strange, long-necked notoungulates roamed the swampy, early Andean coast of South America. As the mountains were uplifted, the primitive puyas adapted to keep pace with climatic changes until they can now survive at up to about 13,000 feet. They grow 30 feet high and are reputedly the largest herbs in the world. In botanical terms, herbs are those plants without central wood structures and the puya belongs to the family *Bromeliaceae*, which includes such diverse species as the pineapple and the straggly, grey-green, epiphytic Spanish moss that adorns the warm temperate and tropical forests of the Americas. According to the American botanist Mulford Foster, a specialist in bromeliads, a mature puya may be as much as a hundred years old.

We left the Land Rover in Catac and walked up the steep east slopes of the Callejón into the Cordillera Blanca, climbing at first along a well-used Quechua trail. We passed a few mud houses and small, unfenced fields. Even down in the valley any exertion had left us slightly breathless, and now we were tiring rapidly. The barrel-chested Indian took short, even steps and maintained a steady pace. It seemed that he could keep going all day, although like us he must have felt warm, because he removed his thin, brown poncho, folded it and carried it over his shoulder. It took us nearly three hours to reach the lower edge of the *puna* at about 12,000 feet, by which time we were thirsty and badly in need of rest. We halted where the trail crossed a mountain stream, and sat on large grey boulders, taking stock of our surroundings.

The snows were still some miles ahead and more than 4,000 feet above us. The towering mountainside appeared dark grey, but the ground all around us was yellow with its covering of tough grasses. These were various species of Festuca and Calamagrostis, known collectively as *ichú* by the Indians. The grasses were just over a foot high and at first sight appeared continuous; but when we looked more closely we found that they grew in clumps with stony ground between. On some slopes the *ichú* clumps were close together, but elsewhere, probably where the ground was drier, they were more widely spaced and the hillside appeared greyer.

The leaves of *ichú* grasses are needle-like to reduce the surface area presented to the dry air and so cut water loss to a minimum. The outer leaves remain on the plant long after they have died, thus protecting the younger leaves that grow in the centre and giving the *ichú* grasses their characteristic yellow colour. The *ichú* clumps also provide shelter from frost and sun for smaller, broad-leaved plants growing in between. Many of these plants were mountain species of *Compositae*, the group which includes dandelions and thistles. They grew close to the ground, their leaves flattened and often almost lost under the cover of the grass.

We rested for half an hour. It was a fine, windless day, and once our breathing had returned to normal we were struck by how clearly the trickle of water over stones rang from the stream. There was no other sound. The yellow *puna*, backed by the sterile white snow and the deep blue sky, seemed empty.

Refreshed, we started to climb again, following the Indian who stepped ahead like a mountain goat.

"Oiga, Quispe!" I shouted (the local equivalent of Hey, George!). "Cut the pace a bit, or we'll be condor bait."

He laughed and slowed down. I stopped for a moment, breathless. My heart was throwing itself around inside my chest like a piece of loose machinery. When we started again, Marion and I went more slowly and soon we had once more fallen well behind our guide. The exertion began to create an illusion of unreality, perhaps because my footsteps and my painful breathing were now the only sounds I could hear. There seemed no scent of earth, flowers, anything. In fact, there probably were smells, but I could no longer detect them. In the *puna*, sensitivity to smell is diminished because the molecules that constitute most aromas are large and heavy, and so not easily carried in the thin air. Without sound or smell, I felt I was losing touch with the world about me.

I was about 30 yards behind the Indian when the ground at my feet exploded with a whirring of feathers. A large, buff-brown bird flew up. I jumped backwards, momentarily off balance. The bird's short wings beat strongly yet lifted it only a foot or two off the ground. The Indian saw it. As if by reflex action, he threw his poncho from his shoulder and chased the bird on its low, downhill flight. It settled somewhere, instantly camouflaged, and he looked around for it. Then it flew off again. I followed the chase for a short while, keen to see the bird at close quarters if the Indian could catch it, but I was soon out of breath and had to sit down. A hundred feet or more below, the Indian finally gave up his chase and walked back up the hillside to rejoin me.

"Un perdiz, señor," he said without a gasp. A partridge.

In spite of its local name, probably given by the first Spanish settlers because of its outward similarity to a European partridge, the bird was one of the tinamou family, a group of poor fliers probably more closely related to the flightless rheas of the *pampas* and plains than to partridges. One reason why mountain species of tinamou (*Nothoprocta pentlandii* and *ornata*) survive in the *puna* is that the chicks can withstand cold of a severity fatal to other species. The Indian led me back up the hill to the point where Marion was waiting for us, and within a few minutes he had found a crude nest between the clumps of grasses. It was loosely made with pieces of grass and lined for softness with a little down. In it were four eggs about half the size of a chicken's, shiny like fine porcelain and coloured a rich, chocolate brown.

After an hour, we skirted a low shoulder and turned into a small valley. On the steep rocky slopes, in sharp contrast to the monotonous

The ornate tinamou—a species ornate only by comparison with its drab lowland cousins—blunders from behind a clump of ichú grass in the high Andean puna. This primitive bird rarely flies except when in panic, and then only for short distances.

cover of *ichú*, stood a thin forest of tall, slender plants. They were the giant puyas. Several were between 20 and 25 feet high, and I approached one of the tallest for a closer look. The main part of the plant was a huge spiky globe of thick leaves tapering to sharp points, all sticking outwards like so many fixed bayonets. This massive globe-shaped rosette was supported on a central, scaly trunk and stood ten feet tall. Above that, like a candle fixed in a Christmas decoration, rose a stalk about one foot thick and 15 feet high, covered with flowers. I scrambled up on to a large rock on the steep hillside and found myself on a level with the flowers. They grew on stubby branches no more than a few inches long, which were wound in a spiral round the centre of the stalk. I could see flowers in different stages of development. Some were faded and dying. Many hundreds were in bloom, their petals of palest, green-yellow forming a shallow trumpet. Where the flowers had not opened, the tiny branches were clothed with buds like unhusked ears of wheat.

For most of its long life, a puya consists merely of the spiky, globe-shaped rosette, and I could see many of these younger plants on the hillside. The puya sends up a flowering stalk only once and then dies. There can be as many as 8,000 flowers on a single stalk, and from them come several million tiny winged seeds with which the aged plant makes a grand, last effort to ensure the propagation of the species.

Flowering puyas are as conspicuous as trees in the bleak, open *puna*, and like trees, they attract a variety of wildlife. I had been motionless for a few minutes when I heard the soft vibrations of hummingbird wings. Two Andean hillstars, the commonest hummingbird species of the highlands, came to the puya—two-inch darts of brilliant green visiting one flower after another. My presence did not seem to bother them. They completed their search for nectar, and disappeared.

I returned to the base of the plant to see if I could detect any signs of life among the leaves. Between the tough sabres I could make out a mass of greyish-brown feathers, and I tried parting the leaves with my camera tripod. The sharp points of the leaves and the vicious hooks along their edges cut my hands and snagged my cotton anorak, but after several jabs I managed to dislodge the corpse of a small bird. In its haste to hide from a mountain hawk or some other bird of prey, it had flown among the spines and impaled itself on one of the sharp points. As I searched deeper between the leaves, I disturbed a small dove. It flew out and almost hit me. Holding the stiff leaves down, I could see its nest made from scraps of *ichú* grass and pieces of dried puya. In several other puyas I found dead birds and long-deserted nests.

PUYA WITH STALK TOWERING ABOVE PUNA GRASSES

Slow-Growing Giant of the Puna

An anomaly of the Bolivian and Peruvian heights called the *puna* is the *Puya raimondii*, which seems to defy the laws of survival in its desolate habitat. It towers 30 feet into the air on mountainsides where most other plants have low, compact shapes more obviously adapted to the restricted growing season and the destructive power of the winds. The reasons for this plant's peculiarity are not clear but some botanists believe that the puya is a relict species that grew in this section of the Andes when the land was flat and marshy, and adapted to higher altitudes as the mountains formed.

The puya takes about a hundred years to reach maturity. The base grows slowly into a sturdy trunk, some two feet thick, topped with a rosette of knife-edged leaves, each leaf four feet long and studded with barbs. These fend off most browsing animals, and the spaces between the sharp leaves provide shelter for some prudent birds.

When the plant reaches maturity, it sends up a bloom stalk with astonishing speed. Within three months —a fraction of the plant's total growing time—the stalk is 13 to 20 feet high, composed of some 600 branches curved around a central spike.

Whitish-green flowers bloom and wither on this stalk, producing millions of seeds for dispersal by the wind. The plant then dies but the rosette continues to provide a shelter and a snare for decades more.

HUMMINGBIRD VISITING DYING PUYA BLOOMS

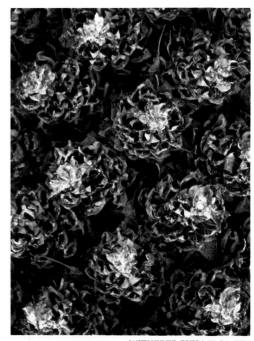

WITHERED PUYA FLOWERS

A NESTING BLACK-WINGED SPINETAIL IN THE ROSETTE

DEFENCES OF THE PUYA LEAVES

During the week we spent in the *puna*, we saw only one other plant type of any size—small trees ten to 15 feet high growing in a narrow, glaciated valley. These were *kenua* or Polylepis trees, which belong to the *Rosaceae* family, a large group including the common strawberry and apple. They are almost the only trees of the cold, high Andes. Their foliage was thin and silvery, covered with silky hairs, and their twisted trunks were protected from the cold by grey-brown scales of bark which came away in my hand like old parchment.

With so little plant cover, animals have to find other forms of shelter. White-winged finches spend the night in great dormitories of 200 or more, huddled beneath rocks and cutting down their heat loss by using one another as insulation. Several species of birds build underground tunnels, and we were on the lookout for banks and hillsides that seemed suitable for avian excavation. We saw nothing until amost the end of the week. Then, in a wide expanse of *ichú*-covered *puna*, I sighted what looked like a group of woodpeckers. One bird was closer than the rest, and when I saw its long, sharp, cone-shaped beak and dull yellow, dark-banded plumage, I recognized it as an Andean flicker, well adapted by its tunnel-building habit to the treeless world of the *puna*.

The birds flew away fast and low over the *ichú*, but I followed their direction and rediscovered some of them on a steep bank above a dried stream bed. One bird was busily enlarging a hole in the bank—the entrance to its burrow. Hoping not to be noticed I wriggled carefully forward on my stomach, trying to ignore the sharp *ichú* which penetrated my clothing. The flicker stopped digging for a moment or two, looked warily about and then settled down to its work again. In the stillness of the narrow gully, the hard sound of the pick-like beak hitting the dry mud continued purposefully.

Suddenly, without any apparent reason, the bird left the hole and flew out into the *puna*. It was just about to rise when I heard a soft rustling in the air above me. I sensed a large presence and looked up. Barely 20 feet above, or so it seemed, was an adult condor, its impressive wing-span of perhaps eight feet almost blocking out the blue sky. The huge vulture had clearly come to investigate me, and although I knew it would not land unless it was sure I was carrion, its close proximity was frightening. There is something evil about this scraggy bird, with its repulsive, naked-wattled head, viciously hooked beak and loose, turkey-like wrinkles about the eyes. The condor rose towards the snowy face of the mountains, its upturned wingtip feathers like a row of spread fingers. Then, almost in slow motion, it banked, displaying the white

A young male Andean condor glides effortlessly, its wing-tip feathers splayed out to form independent aero-foils that give increased lift.

markings on the upper surface of its dull, black body, and came towards me again. As it approached I could see its bare head and neck turned inquisitively in my direction. I knew it was studying me closely, for condors have excellent eyesight.

This faculty was pinpointed as early as the 1830s by the young Charles Darwin, who conducted an experiment to find out whether condors located food by smell or by sight. He tried wrapping putrid meat in paper and offered the foul smelling package to some trapped condors. They did not react until he tore off the paper, so Darwin deduced that sight and not smell was the vital sense. It has since been established that, like other birds, the condor's sense of smell is weak.

This condor must have seen me make some slight movement, for on its second pass it came over twice as high as before and rose away towards the mountain on an upcurrent of air. I stood up and watched through binoculars as it sailed on magnificently with hardly a wing movement until it was a speck lost in the icy depths of the cordillera.

Ears alert, a mountain viscacha stands watch over a rocky retreat more than 13,000 feet up in the Andean puna. One of the world's highest living mammals, it can range to about 16,000 feet so long as grass, moss and lichen are available at that height for food.

One *puna* animal we had not seen so far was the vicuña. Indeed, we had not even seen any vicuña tracks or droppings. These delicately limbed, graceful, cameloid animals are found only in a few parts of the central Andes, and we decided to leave the Cordillera Blanca and search for them in a more southerly part of the Western Cordillera where we had had information of their whereabouts.

When the Spanish arrived in South America, the distribution of vicuña extended from Ecuador to northern Argentina. This thriving population was at least partly the achievement of the royal line of Incas, the rulers of the highland Indians, who strictly controlled the hunting of the vicuña. The animals were protected until well after the wars of independence in the 1820s and 1830s. Their numbers diminished only gradually up to the Second World War, but the decline accelerated after the war and the animal was hunted almost to extinction. The Peruvian government established a reserve at Pampas Galeras, above Nazca, and a Peruvian conservationist, Felipe Benavides, crusaded internationally for laws to prevent the importation of vicuña skins into the United States and Europe. Gradually, the decline was halted, and between 1967 and 1974 the numbers of vicuña on the reserve increased from fewer than 2,000 to more than 12,500. It was not quite a zoo, we were told, but there were plenty of vicuña there. Since we wanted to see vicuña in truly wild conditions, we set off to search for them in a desolate stretch of *puna* in the Western Cordillera south of the volcano Ubinas.

We drove into the mountains by Land Rover with Mario, a Peruvian biologist, as our guide. At 14,000 feet we entered a wide, open expanse of semi-desert *puna*, where mile after mile of *ichú* grass spread across rolling hills. We were following a corrugated dirt road, and after a while Mario decided that we should branch away from it cross-country and search for *vicuña* among gullies he had previously seen from the air.

For the first hour we made little headway; the only possible route took us in a series of tight loops through the narrow gullies. In places, the road was blocked by dense masses of bright, yellow-green *tola* bushes, growing well perhaps because there was water in the gravel of stream beds. We began to climb steeply through sand and clumps of *ichú*. Rocks appeared, and we had to find a way through cliffs of pink ignimbrite, a reminder that we were not far from great volcanoes.

In the early afternoon we found a long sandy slope ascending gently between low boulders. The arid *puna* seemed almost uninhabited. The only living creatures we had seen all day were buff-grey ground doves which flew up from *tola* bushes as the Land Rover brushed by. There was no sign of any Andean Indians, no relic of any human settlement. The slope led to a wide gully flanked by low cliffs. This too seemed empty of animals and there were no plants apart from a few scraggy *tola* bushes. Uncertain of which way to go, we stopped.

I took the opportunity to do a little exploring. As I walked across the sand, a quick movement caught my eye. I had been seen first, and all I glimpsed was a grey tail and a pair of hind legs scurrying for cover. Then I heard a whistle. I turned quickly, half expecting an ambush, and saw what had made the sound. On the other side of the gully the rocks were alive with furry rodents which looked like a cross between squirrels and large rabbits. In response to the warning signal given by one of their number, they all slipped out of sight among the rocks. I recognized them as mountain viscachas, relatives of the almost extinct chinchilla. I scrambled among the bare rocks and saw that the ground was littered with tiny, ovoid, faecal pellets.

The mountain viscacha is one of the world's highest living mammals and is found up to 16,000 feet in the Andes, living in colonies of around 80 individuals that exploit natural shelters among the rocks as protection against the cold. Like so many *puna* animals, the mountain viscacha is a small member of its family, weighing about three and a half pounds compared with the 15 pounds of the viscachas of the grassy lowlands east of the central Andes. A viscacha may reproduce two or three times a year, but usually only a single offspring is born, for with the scarcity of food in the *puna*, the mother is unlikely to be capable of suckling more.

Looking at the sand between some *tola* bushes, I noticed among the viscacha tracks two rows of small marks, like large dots arranged in parallel lines. They were about an inch and a half apart, and between them was a single, unbroken line. A lizard had passed through here, dragging its tail. When I examined a larger area, I found scores of these trails, but for a while no lizards. Then a sudden movement gave one away. The little reptile was about six inches long and had been resting on the sand. Its pale, lime green skin, darkened with blotches of brown, formed a perfect camouflage in the shadows cast by the *tola* bushes, and had it remained motionless I would not have seen it.

The Andean lizard is expert at raising its body temperature against the *puna* cold. During the night it shelters in its burrow. At the first sign of sun, it pokes out its nose and tests the air. What little heat there is so early in the morning warms the blood vessels in its sinuses, and from here warmed blood begins to circulate round the rest of its body. When the lizard feels able to, it crawls out of its burrow on to some material that will absorb the sun's heat quickly—often black, decaying vegetation—and, flattening itself, begins to bask. This spread-out position increases the surface area of its body, allowing the lizard to absorb the heat of the sun and of the vegetation underneath. It also reduces the lizard's thickness so that the heat has less tissue to penetrate. Thus although the air temperature may remain at around freezing, the lizard may heat its body to more than 85° F. and become active.

Where the *puna* was clothed with *ichú* I would not have seen so many tracks, and our pause in the sandy gully was proving most productive. Then came the greatest stroke of luck so far. In one place the sand was disturbed by a confusion of indistinct depressions obviously caused by a group of large mammals. Some vicuña had been here. We set out again in the Land Rover, following the tracks through a constricted part of the gully until we came to a steep, boulder-strewn slope. Before attempting to climb it, we off-loaded some heavy gear, including spare fuel. The engine had become sluggish at the high altitude, but now we found it surprisingly lively in second gear, considering that we were at 14,000 feet. Marion drove up the slope, gaining a hundred feet, and we came to the edge of a dry, brown plateau. The sandy earth was almost bare of vegetation, and the vicuña trail led across it to some rocks a good two or three miles away.

The blue sky became overcast as we drove across the plateau, giving the place a sombre gloom. Just as we approached the rocks on the far side, three vicuña broke cover and sped away to our right like golden bullets. Marion turned the Land Rover to follow. This was our chance to see the animals at close quarters. The engine raced in low gear and on the level ground we were not far behind. We were doing over 20 m.p.h. in second, bumping over rocks and scattered clumps of *ichú*, slowing suddenly in soft sand, hitting hard ground again with a sharp scream from the engine as it whined back up to maximum revs. We stayed less than a hundred yards behind the vicuña. They bounded along so effortlessly that at times they looked as though they were flying above the surface of the sandy uplands. Their speed was certainly enough to protect them from mountain cats, pumas and foxes.

The three we were following were not quite fully grown and measured about three feet to the shoulder. They were young males recently exiled from their family group, according to the norms of vicuña society. The family groups, each led by a single male, have well-defined territories which provide protected feeding areas for the females and their young, while the single males scratch a living on the margins until the strongest among them are able to establish family groups of their own. The stamina of these three young vicuña was astonishing. For several miles they ran steadily, necks outstretched, ears sharply back. At times we were close enough to see small clouds of dust spurting from the ground beneath their feet. When we reached a softer part of the plateau, our engine could not compete with them and they easily moved away. Suddenly they disappeared. Then, one by one, they emerged from the plateau as though pulled from a conjuror's hat. We soon found the explanation for their vanishing trick. A deep gully with soft sandy sides separated us from the sprinting animals. We stopped at its edge and watched them disappear into the rocky wasteland of the *puna*. They were light, fleet and wonderfully elegant creatures. I would not have chosen to see them anywhere but in this immense, dusty landscape to which—like the other animals of the *puna*—they were so finely adapted.

A Spectrum of Wildlife

From the Pacific coast of Peru, inland across the narrow hump of the central Andes, and down the eastern slopes—in a band only 350 miles wide—five different environments, ranging from desert to jungle, support an amazing cross-section of Andean wildlife.

On the rocky, coastal headlands up to 30 million cormorants, gannets and pelicans gather in raucous colonies, attracted by massive shoals of anchovy that thrive on plankton in the cold waters of the Humboldt Current. In some years, disaster strikes the colonies when a warm current, the Niño, displaces the cold waters, driving away the plankton and the fish. Millions of birds, unable to find new feeding grounds, die of hunger and exhaustion, leaving the Peruvian coast thickly littered with their carcasses.

Behind the Peruvian coast, the low, western slopes of the Andes are desert, broken by an occasional oasis of valley and river. Only small animals live here, such as rodents, foxes and lizards which can hide under boulders from the sun and can derive most of their water from the food they eat.

Across the top of the range, between 12,000 and 15,000 feet, lies the *puna*, or wide open grassland. Here there are a few large animals like the puma and two wild members of the camel family, the guanaco and the vicuña. The vicuña's soft wool is more than three times finer than merino wool, and its enlarged heart and lungs adapt it to life in the rarefied air of the high mountains.

Beyond the lifeless, snow-capped peaks and down the eastern slopes of the Andes lies the cloud forest, a zone of dripping vegetation between about 11,000 and 3,000 feet. In this forest live an abundance of insects, birds and, on the lower slopes, small mammals such as spider monkeys, peccaries and margays. One of the best camouflaged small mammals is the silky anteater, a shy, golden-furred, arboreal creature that often conceals itself among the fleecy pods of the silk-cotton tree.

The greatest variety of life is found in the last of the five zones: the lowland tropical rain forest. Here live jaguars, water-loving tapirs, capybaras and huge snakes like the anaconda and bushmaster. Through the tree-tops fly pairs of macaws, the largest birds of the parrot family, enlivening the jungle with their loud screeches and their vivid, red, green and blue plumage.

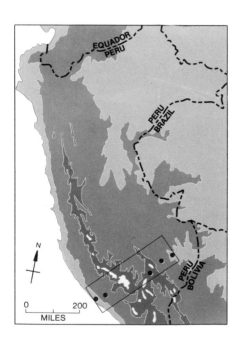

The boxed area on the above map encloses the 350-mile-wide band of southern Peru containing the variety of animals shown in this essay. Each red dot, from left to right, indicates the approximate altitude at which each animal lives. From sea level to 750 feet is light brown; 750 to 12.000 feet, medium brown; 12,000 to 14,000 feet, dark brown; peaks of 14,000 feet and over, yellow.

Brown pelicans, satiated with fish caught in the off-shore waters, rest on the guano-spattered rocks of the Peruvian coast.

In the western foothills of the Peruvian
Andes (right), loose sand and dark,
weathered rock create a landscape
hostile to life. Among the few animals
to survive here are reptiles like the
12-inch tropidurine lizard (above),
a member of the iguana family. After
hiding from the midday sun, it has
emerged to hunt insects and spiders and
to rasp at cacti and other desert plants.

In the wide open puna of the high
Andes (right), a line of vicuña
disappears over a ridge. Seldom seen
except from a distance, vicuña travel
in groups usually consisting of
either a male leader (above) and up to
15 females, or a number of leaderless
males. In spite of the low atmospheric
pressure and lack of oxygen, vicuña
can run at 30 miles an hour,
vanishing into the highland wastes.

On the eastern slopes of the Andes,
the draped thickets and tangled
undergrowth of the lower cloud forest
(right) harbour many small animals
including the silky anteater. This
elusive, largely nocturnal creature, seen
here sleepily scratching its snout,
grips its perch with two large hind
claws, coiling its prehensile tail around
a lower branch for stability.

Above the dense canopy of the lowland jungle (right), a small party of bright-red macaws fly in search of nuts and fruit in the trees. These birds have a powerful beak, clearly displayed by the red and green macaw above, with which they can even crack Brazil nuts.

5/ Savage Waters

Between the ranges in these mountains to the west and to the east, far down, a great white swath crawled like a serpent . . . It was the Marañon, a river as great as the mountains and the jungle. CIRO ALEGRIA/ THE GOLDEN SERPENT

Inland from the dry Western Cordillera, the Andes of southern Peru are violently folded and contorted for 150 miles before dropping steeply into the Amazon rain forest. North of the volcano Ubinas, parallel ranges are entangled with other, transverse ranges to form gigantic knotted mountains whose summits are sheathed in ice. Somewhere among these Peruvian *sierras*—geographers cannot agree precisely where—lies the source of the world's greatest river, the Amazon. The longest tributary of the Amazon, the Apurímac, begins in the snows and flows northwards. Within 500 miles it cuts through deep canyons and plunges more than 15,000 feet down the precipitous eastern slopes, its name changing four times. First it becomes the Ene, then the Tambo; then it joins the Urubamba to form the Ucayali; and hundreds of miles farther north, the Ucayali joins the Marañon to become, finally, the Amazon.

By the time it leaves the Andes, the river is twice as wide as the Rhine, although it still has more than 3,000 miles to flow to the Atlantic. When it reaches the sea, it carries 20 per cent of the world's fresh water; and much of it comes from the Andes. The volume of water is massively swollen by the run-off from thousands of square miles of soaking forest on the eastern side of the *sierras*—the *montaña* as it is known in Peru. Heavy rainfall is created as the trade winds, blowing in from the Atlantic over the Amazon basin, meet the *montaña*. As they rise up the steep slope, they become progressively cooler in the higher altitudes.

The moisture with which they are laden condenses, forming cloud and engulfing the *montaña* from around 2,000 to 3,000 feet upwards—the figures vary significantly from place to place. Higher up, the density of cloud increases, reaching its maximum between 9,500 and 11,500 feet, where rainfall is 80 inches a year or more. This zone of soaking vegetation may loosely be described as cloud forest, a term which encompasses various levels of montane and subtropical forest. Here the tree canopy is lower than in the tropical rain forest, or *selva*, of the lowlands and there is sufficient light to sustain a thick tangle of undergrowth. Higher still, the rains become lighter and the cold more intense. A zone of dwarf forest replaces the cloud forest, then gives way to the grass of the cold *puna* and finally to the permanent snow of the peaks.

During the rainy season between February and April, millions of tons of water flood down the eastern slopes, often causing gigantic landslides. The rains sweeping over the *sierras* soak the high *puna* of the mountain interior, which is dry for six months or more every year. In April when the heavy grey cloud is at last broken by patches of blue, the landscape is brightened by tall, white masses of cumulus which signal the retreat of the last shower as they disappear beyond the cordilleras.

I had always wished to gauge for myself the volume of water pouring down from the Andes to feed the Amazon. The chance came when I made a journey in the rainy season, starting from the lowland rain forest and travelling by canoe up some of the major tributaries into the eastern cordilleras. Having struggled against the flood at the base of the mountains, it was my intention to trace the waters back to the high *puna* and the snows of the Peruvian *sierras* near the high mountain town of Cailloma, where the Amazon is now generally considered, after centuries of argument, to have its source. Almost certainly the longest tributary of the Amazon is the Apurímac, not another old favourite, the Marañon; but exactly where the Apurímac rises is by no means clear. Source hunters have found likely spots all around Cailloma. One place officially supported by the National Geographical Society, is Mount Huagra, just north of the town, but most of the possible locations are in the Cordillera de Chila, a crescent-shaped range encircling Cailloma to the south-west, and it was to these mountains that I decided to travel so I could see, if not the source itself, at least the type of country in which the world's greatest river rises.

I have watched both the Marañon and the Apurímac in flood: thick, brown torrents ripping at their banks, carrying away millions of tons of debris and toppling giant trees into the current. It may seem foolhardy to venture out into rivers such as these; yet some boatmen argue that high water is safest, for that is when many of the most dangerous rocks and rapids are submerged.

I made this journey without Marion, flying in a light aircraft to a forest clearing beside the river Tambo, one of the lower reaches of the Apurimac. Here I had arranged to go upriver with an experienced boat-man, Hugo, and his Indian helper who were making a trading expedition in a 30-foot canoe powered by a large outboard motor. Hugo was from Quillabamba, a small town in the eastern Andes not far from Cusco. He had spent much of his life on fast-flowing rivers, and we were going to need every bit of his skill. Ahead of us was a notorious stretch of the Ene, where the water, in its final tussle with the mountains, is flung about in a maelstrom of rapids and whirlpools.

For two days the large outboard droned steadily as we made our way upriver. All around us was wild, virtually uninhabited rain forest, and ahead of us the course of the river led relentlessly upwards. In places I was sure I could see the surface of the water sloping uphill, and the motor had to push hard against a strong current. The unbroken line of trees on the banks and the gluey-smooth surface of the brown water were mesmeric. In places the Tambo was nearly 300 yards wide and dangerously near the top of its banks. I had been here once before in the dry season, when the banks were ten to 20 feet high, so I knew we were afloat on an immense flood.

Rocks began to appear in the river on the third day, when I caught my first glimpse of the Andes. Through a thin veil of mist, distant tree-covered buttresses interrupted the monotony of the jungle, soaring high above the tree canopy like a green wall. For centuries the Indians of the *selva* have regarded these mountains as the end of their world, believing that some demon of the heights governs the turbulent water rushing down the canyons. I could see why. The river had now assumed a thoroughly malevolent form: submerged trees and rocks, then rapids, some gentle and some noisy. We met these hazards one by one.

Hugo sat in the bow, watching the river intently. He could interpret the eddies and ripples on the surface of the water and knew where sub-merged obstructions might capsize the canoe or foul the propeller. His helper, the *motorista*, held the tiller and took signals. Together they conducted a classic piece of river navigation. A slight nod or a hand sign from Hugo would direct our course around something hidden just below the surface. But even with his formidable command of the river, Hugo occasionally warned the man at the stern a second too late. At one nasty moment, the long canoe glanced against the bough of a submerged tree and tipped dangerously. "Derecha, derecha!" Hugo shouted above the noise of the engine, and we turned sharply to the right.

We had reached a bend where the river was deflected by a rocky spur. At the curve, millions of tons of water were forced to change direction sharply and the outer bank was rapidly being undermined. Sometimes, in places like this, hundreds of yards of undermined earth and forest collapse with a thunderous roar into the swollen river. The trees are swept downstream until they are caught in some shallow spot. When the water subsides, the skeletons of such trees are revealed; but now their branches were invisible beneath the surface. They were some of the obstacles that Hugo was looking out for. But they were nothing compared to the one that now confronted us.

At the bend, where the water almost stopped before rushing on downhill, I saw a *remolino*, a huge whirlpool. Its gaping vortex was nearly 30 yards across. Even a hundred yards from the edge we could feel the powerful tug of the current.

The engine was already flat out and we had nothing in reserve but Hugo's skill as a riverman. With a flick of his hand, he confidently directed the *motorista* close to the opposite bank. We moved within touching distance of the forest. Here the water was quieter and the motor began to gain on the current.

As we slowly edged past the slurping hole, a giant tree from farther upstream was swept into the spinning current. The river had stripped the boughs of leaves and bark, leaving them a pale, creamy brown. Little by little the scoured skeleton was sucked downwards until the tree was standing nearly upright and only the topmost branches were above the water. As much as 30 feet of timber had been dragged beneath the surface. Just as I thought the tree was about to vanish, the force below released its hold and the forest giant was cast like a twig to the outer rim of the vortex, to begin another slow revolution from which there seemed to be no escape. We moved away from the danger, and looking back I saw the whole sequence repeated once more.

In another hour we had reached the beginning of the river Ene, between the lowest foothills of the Andes, and the red banks gave way to cliffs and boulders. The river rushed past us. Even with the motor at full bore, we were often stationary in the current and at times even forced downstream. Hugo was expert at spotting useful back-currents into which we edged, and usually they were enough to help the canoe forward. We refuelled frequently from the 40-gallon drums which were settled in the middle of the canoe and lunched on dry biscuits and a sweet mixture of cooked bananas and tinned milk—regular river fare. I leant over the side and scooped an enamel cupful of water. Hugo said it was perfectly safe to drink.

I took two or three gulps of the turbid Ene and found its flavour delicious, with no trace of the purifying agents one gets so accustomed to in the city. Although the taste was earthy, it was not harsh or alkaline like some spring waters, and it reminded me of the softness of newly fallen rain. Of course, that is what most of it was. Heavy rain was falling along a 1,500-mile front ahead of us, feeding thousands of Amazon tributaries from north of the Equator almost to the Tropic of Capricorn. Dense black rain clouds hung in the mountains. Water cascaded down the cliffs beside the river. Every tributary we passed was full. It was as if some gigantic tub full of water had been sluiced over the Andes.

We ventured on for several hours, encountering more rapids and whirlpools and now making little progress against the Ene. In places where large boulders obstructed the current, the water rushed past with such force that it did not quite meet on the other side, and deep pits were formed. Sometimes I felt I could see right down to the river bed. Eventually we ceased to make headway, and there was no alternative but to turn back. We camped overnight at the edge of the forest, and next day we made a quick run downstream again.

I had fulfilled my ambition to experience the Amazon's source tributaries in flood; but having failed to get farther than the foothills, I continued my journey by other means in order to see where so much of this water originated, three miles up in the mountains near Cailloma. From the jungle airstrip beside the river Tambo, I flew to a roadhead in the eastern Andes and from there I travelled in an assortment of local buses and trucks, making my way along the middle section of the Apurímac gorge and finally reaching Cailloma, high in the *puna*. This journey turned out to be almost as dangerous as the river expedition, and I was held up three times by landslides. In one place at the edge of the Apurímac gorge, a 50-foot section of the road had fallen 200 feet into the river. When I peered over the broken edge of the gash, stones and water were still tumbling down.

"Señor, señor, a miracle!"

"Dios mio! Dios mio!"

An excited hubbub rose from the small crowd which had gathered. It

Two tiers of a waterfall plummet 1,500 feet down the sheer face of the remote Cutibireni plateau in central Peru. Like more than one hundred other falls that drain the plateau, it has an intermittent flow. Its greatest volume is after the downpours of February-April, dwindling to a streamlet when there is no rain.

did seem like a miracle. One truck driver had just managed to stop as the road collapsed in front of him.

After three days of hard travelling, I reached Cailloma. There I found a truck driver who agreed to take me at least some of the 30 miles across the *puna* to the snow-capped Cordillera de Chila, the likeliest location of the source of the Amazon. It was a cold, rainy day when we left the town. The streets were muddy and the few people who were about kept close to the adobe walls of the houses to avoid the rain. We followed a rough track upwards across the *puna*, which in the dry season would have been yellow and parched, but now was dark and boggy. The steep uphill sections of the track were deeply scored with ruts which were always a disastrous axle-width apart, so the large differential between the rear wheels snagged and bumped on the hump in the middle. Part of the trouble was that the truck was empty and there was not enough weight to hold the rear wheels to the ground. Frequently the tyres lost their grip and spun frantically.

Drivers in the Andes are used to conditions like these, and they generally calculate distance in hours rather than miles or kilometres. If you ask a driver how far it is to a certain place and he replies— probably with a shrug—that it will take many hours and cost a lot, he means that he does not want to go there. I risked asking this driver how far it was to the end of the road. "It depends," he replied with the tell-tale shrug, looking at the track ahead. "But . . ." It was an obliging qualification, but before he could mention a price, I announced that I would continue on foot.

It was raining hard and the hills all around were hidden by heavy cloud which cut the horizon to a bare hundred yards. The clumps of normally tough *ichú* grass were soft and bedraggled, and set like stepping stones in a giant quagmire. The cold damp air penetrated my thick clothing and seemed to pass straight through me. Each step was a squelchy effort. I soon found myself stumbling breathlessly from òne grassy clump to another, and recalled that I was already between 15,000 and 16,000 feet—more than three miles above sea level.

I understood now the difficulties of pinpointing the source of the Amazon. The ground was so sodden and filled with rivulets that it was impossible to distinguish one from another. Even in the dry season, there would be plenty of little streams to choose from, all leading down from the snows and glaciers of the cordilleras.

When I reached the top of a steep slope, I found that the rain was

Where the edge of a puna stream has frozen, fragments of water weed and gravel are encapsulated in ribbons of ice. The stream freezes over daily, beginning in the late afternoon. In the thaw the following morning, the ribbons of ice melt last and are marooned along the banks when the body of the stream resumes its regular flow.

mixed with sleet and snow. Only a short distance beyond me, hidden in cloud, was the continental divide, the snow-capped ridge from which streams on this side flowed through canyons and forests to the Amazon, and those on the far side led down the dry western slopes of the Andes to the Pacific. This icy watershed between east and west is the only point that can be identified without argument as the end of the Amazon. If I climbed any farther, I would soon be in snow. I was soaked to the skin and numb with cold, and felt I already had more than a sufficient idea of the headwater country. I decided to call it a day.

I had one more ambition to fulfil in the Peruvian *sierras*. I wanted to explore one of the high glaciers whose waters join the torrential rains and swell the tributaries of the Amazon. In the Andean winter month of May, when the storms had ceased and the curtain of cloud had been drawn back, I returned to the area. This time, instead of revisiting the Cordillera de Chila, I made for a much more dramatic range of mountains, the Cordillera de Carabaya, a hundred miles farther east. This is part of the great line of eastern cordilleras which overlook the Amazon basin. To the north is the Cordillera de Vilcanota and to the south the Nudo de Apolobamba and the Cordillera Real, bordering Lake Titicaca. Many peaks in these cordilleras are 18,000 feet high, and some, like Ananea in the Nudo de Apolobamba, reach 19,000 feet.

I managed to get a lift in a truck, and we approached the range from the west along a dusty road at not much more than 12,500 feet. The mountains towered ahead of us, their snowfields and glaciers crisp in the winter sunshine. Some tongues of ice reached below the line of permanent snow, shining a wonderfully pale, almost phosphorescent blue. When we were close, I asked the truck driver to drop me off at a small Indian settlement that I could use as a base for a hike into one of the glacial valleys. I arranged to stay the night on a rough bench in a roadside hut where an Indian and his wife made a living by providing food and shelter for travellers and keeping a few llamas.

The next morning I set out towards the cordillera in the company of a local Indian who was travelling to a higher settlement to visit a relative. One starts early in the mountains to take best advantage of the daylight and as we were within the tropics where the length of day does not vary much between summer and winter, I knew I had a good 12 hours before dark. The climb was gentle at first, but then steepened sharply. Some time after parting from the Indian, I entered a high mountain valley and began to walk upstream along the clear, ice-cold rivulet that was cutting a shallow trench in the valley floor, exposing grey-brown mud and smooth, rounded stones left behind from glacial erosion long ago. In some places enormous, worn boulders blocked my path, and I had to skirt them by climbing over piles of rocky debris.

Soon I was well above 16,000 feet. One side of the valley was in heavy shadow, and out of the sunlight the air was cold. I kept to the warm side, but I could see that the smooth rocks in the shade were thickly covered with ice. It would not melt for another three months.

As I climbed higher, the ground became steeper, the air thinner, the vegetation sparser. I became breathless and hot and removed my anorak and woollen sweater, stuffing them behind a rock for collection on the way back. I heard the noisy chatter of some Andean parakeets and looked up to see at least 20 of these tiny birds racing across the valley, their brilliant green plumage piercing the austerity of the mountains. The earth was brown, patched with yellow grasses. Here and there the small rosettes of other plants spread over the ground, and there was even the blue of a late Andean lupin. Everything was dry; nearly all the water from the summer storms had long since gone or was locked away in the frozen mountains above. In two months the *puna* had changed from floods to semi-desert.

On the sunny side of the valley I noticed a long ledge standing out on the slope about 30 or 40 feet above the stream. It was clearly a lateral moraine, a mass of rock debris heaped against the side of the valley by the glacier that had once scraped its way through here. I climbed up the steep slope of the moraine, slithering on a thin layer of soft humus which had built up over the centuries, and continued along the top. Looking back down the valley, I could see the remains of an old terminal moraine, where a dam had once been formed from one side to the other by debris brought down in front of the glacier. The ancient river which flowed from the glacier had later cut a way through the obstruction.

Higher still, I found another terminal moraine marking the position reached by the glacier at some more recent stage in its history. This dam was intact, and behind it I discovered a small glacial lake about 100 yards wide by 200 yards long. The water was blue-green, and in the shallow parts it was frozen. Ice enveloped the plants at the water's edge. On a wet, hummocky spot a pair of large Andean geese watched me suspiciously before taking to the air.

Above the lake I could see the retreating glacier in its present position. Its snout lay as little as a hundred feet away in vertical distance but it took me nearly half an hour to reach it, scrambling across the rubble in the bottom of the valley. When I got to the glacier, I was surprised. It was not the pale blue of the glaciers I had seen from below, but grey from the tons of small stones and rocks embedded in it. The snout was 20 feet high and more than one hundred yards wide, bounded on each side by heaps of light grey, fresh moraine debris. At its foot only a trickle of water emerged; in the summer it would grow into a rushing stream. After climbing to one side, I could see the glacier curving away up the high valley it was helping to carve. This and other glaciers like it were slowly wearing away the top of the Andes like giant rasps.

I walked up one side of the glacier, following its lateral moraine. After another half hour, I reached the snow-line and found a series of crevasses in the ice which seemed to invite exploration. Some were deep, but I found one small chamber with walls only two feet apart, and went in a yard or two. I looked back to the narrow entrance, filled by a deep azure sky. A fringe of long, smooth icicles caught the brilliance of the Andean sunlight and cast a shimmer on to the glassy walls. I took off my gloves and touched the ice, the accumulation of sleet and snow from wet seasons many centuries ago. In the frozen chamber there was neither sound nor movement, yet in the few minutes I was there, the glacier must have flowed fractionally downhill.

Less than a mile away, horizontally through the mountain, other glaciers were moving down the precipitous eastern slopes above the Amazon jungle. In some parts of the eastern cordilleras the drops are sheer for thousands of feet, and on the eastern side of Mount Illimani in the Cordillera Real there is an appalling precipice of at least 4,000 feet. When the British mountaineer, Sir Martin Conway, led an expedition up Illimani in the late 19th Century, he looked over this precipice at night. "The blanched snow and black crags appeared dim and ghostly near at hand," he wrote. "But the dim and vague horror of that almost fathomless plunge into the dark gulf at our feet was one of the experiences it has been worth living to know." As I stood in my chamber in the gleaming ice, I made up my mind to move on to the Cordillera Real and see its horrific slopes for myself. I also planned to go a stage farther than Conway and try to descend through the cloud forest, down to the lowland rain forest from which I had started.

NATURE WALK / Descent to the Cloud Forest

WITH PHOTOGRAPHS BY MARION MORRISON

The Bolivian night was bitterly cold. I waited for dawn to break, huddled in sweaters and an anorak at 14,000 feet on a steep eastern face of the Cordillera Real. It was 4 a.m. Only a yard or two in front of me the mountainside slid away into darkness towards the jungles of the Amazon, unseen far below. Behind me, the snowy peaks of the cordillera were just visible against the sky.

I had come over the mountains the day before along a high, snow-covered track with Marion, who was going to take the photographs, and Peter Williams, an English engineer who had long experience of climbing in the Bolivian Andes and knew of a way down the mountain into the cloud forest on the precipitous slopes immediately beneath us. I was the only one awake.

At 4.15 a faint glow lit the horizon beyond the Andean foothills where Bolivia borders Brazil. An hour later a broad gash of crimson spread between distant layers of cloud and rich orange rivers of light flowed towards my lookout. Finally, at 6.15, the sun rose to reveal a dense forest covering the hills and plains to the east. The warmth of the sun bathed the high mountainside and instantly softened the night-chilled air.

As the light strengthened, I saw the country immediately below me for the first time. It was the source valley of the river Challana, whose tumbling waters flow ultimately into the Amazon—not a simple valley, but a deep depression etched with a branching network of hundreds of valleys and streams. Somewhere about 3,000 feet below me was the beginning of the cloud forest.

Cloud forest is an umbrella term for the various levels of mountain rain forest growing between about 11,000 and 3,000 feet. It is a feature of the eastern slopes of the Andes from north of the Equator right down to northern Argentina. The cool, moist atmosphere of the cloud forest, combined with the ample light filtering through the trees, creates a luxuriance of plant growth which is hard to imagine. Tree ferns and carpets of moss grow everywhere, interspersed with richly-coloured flowering plants such as fuchsias, begonias, geraniums and various epiphytic orchids.

As the early morning sunlight crept down and warmed the valleys, moist air began to rise into the cooler atmosphere higher up, condensing first into wisps of fog and then into thick cloud that lapped gently against

CLOUD FOREST CANOPY WITH PURPLE-FLOWERED TIBOUCHINAS

MALE TORRENT DUCK WITH CHICKS

tensity. The path became steeper and suddenly we found ourselves in thick cloud. We saw our first large trees and several huge tree ferns slowly dripping moisture. We had entered the cloud forest.

Rounding a rocky spur, we caught sight of the Challana, a stream five feet wide, tumbling furiously down the mountainside beneath us. It was a perfect place for torrent ducks, and I was not surprised when I spotted the sleek white head and neck of a male duck crossing the stream with a family of chicks.

A Thick Carpet of Plants

The vegetation around us was vigorously green. A thick carpet of mosses and tiny liverworts covered the ground, and the path was overgrown with bushes, among them Oreopanax, a relative of ivy. Some of the trees were podocarps, related

the mountain like a sea of cotton wool. Within half an hour, the valleys were hidden from view.

I walked cautiously to the edge of the precipice and listened. Far below, a waterfall was leaping into the chasm with a rattling of stones. A parrot called raucously. It was time to start, for we planned to descend right through the cloud forest to the lowland rain forest at the bottom, and it would take all day.

I woke the others and we began our descent, following an old Indian trail that led steeply through the puna and dwarf forest clothing the high slopes. After two hours we were enveloped in thin mist which diffused the light with blinding in-

AN OREOPANAX SHRUB

CHUSQUEA, A BAMBOO

Unexpectedly, we reached bare mountainside. A landslide had carried thousands of trees and bushes down an 80-degree slope, almost to the bottom of the valley, leaving a muddy, boulder-strewn scar nearly a hundred yards wide. Above us hung tons of loose rock, and below was an unnerving 300-foot drop. We edged across the bare ground, hardly daring to look up or down.

On the far side we failed to find the trail again. The only obvious pathway now was the course of the Challana, but along the stream the daylight was stronger and the banks were covered with a seemingly impenetrable tangle of plants. We had to cut our way through with machetes. After each cut we slithered downhill as far as the vegetation permitted. Occasionally the stems of a scrambling bamboo called Chusquea

whipped back at us, but most of the shrubs were incredibly soft and some apparently tough tree trunks were so rotten with damp that they fell to pieces at the lightest touch.

We had soon descended a thousand feet, although not quite in the manner we had planned. We were covered with wet leaves, scraps of rotting plant stems and twigs mixed with peaty earth. "This is the ultimate in masochism," commented Peter as he struggled through another bamboo thicket, adding with a grin: "Never again."

In the undergrowth I noticed a beautiful Cleome, or spider flower. The wine-coloured stamens radiating like a spider's legs from its centre were decorated with tiny drops of water, so evenly spaced that they might have been crystal beads on a piece of jewellery.

to conifers and also to the Chilean monkey puzzle tree.

The path kept well above the stream and in one place narrowed to a shelf less than two feet wide, cut from the rock. We realized that we were following a mule track made in early colonial times when the Spanish had searched the area for gold and precious stones and rare spices. The old trail was still clear, so we decided to keep to it for at least the next thousand feet downhill.

By 10 o'clock, the cloud had become so thick that visibility was reduced to a few feet. A strong wind blew the saturated air through the ghostly forest in great billows, brushing the trees with moisture.

SPIDER FLOWER

LYCOPODIUM OR CLUB MOSS

pale green Lycopodium spread across a bank just above me. This primitive creeper, with its many branching stems only a few inches high, is often called a club moss or ground pine and is related to some of the earliest plants, now extinct.

Not far away was a typically Andean plant, the Calceolaria or slipperwort, a member of the same family as foxgloves and snap-dragons. It bore a number of deep yellow flowers just under an inch across whose lower petals formed large, slipper-like pouches. Insects are attracted into these in search of nectar, and when they leave, they carry away pollen which fertilizes plants elsewhere in the forest.

I had never seen a Calceolaria in its native habitat before and I found myself almost mesmerized by the movement of the flimsy pouches as

they swayed in the imperceptible breeze. When I bent down, I noticed a number of minute black beetles on some of the flowers: pests, I imagined, since Calceolarias are usually pollinated by bees and wasps.

With so many plants pressing in on all sides, we were beginning to feel claustrophobic. The dampness, the gloom and the lack of horizon in the cloud forest were overpowering. We had entered a strange half-world, cut off from the clear skies of the Andes above, and far removed from the wide, rain-forested plains below.

Beyond the Point of No Return

We started down the canyon. For a few yards we had to wade up to our chests in cold water, and the smooth rock walls closed in around us. We descended 500 or perhaps a thousand feet before the stream emptied into a large open space.

By now the wind had dropped, but the air was still cold, and I was surprised to find that it was nearly midday. We had reached only about 9,000 feet, and if we wanted to get to the bottom of the cloud forest before dark we would have to hurry. Unfortunately, the next stage of our descent looked even more difficult. Ahead, the stream passed through a narrow canyon with sides 50 feet high. Pools and waterfalls lay along the stream-bed where boulders had been caught between the sheer walls. We would have to wade through the rushing water, but first we decided to have a rest.

All around, the ground was covered with a loose green moss, and a

CALCEOLARIA FLOWERS

FUCHSIA FLOWER

brief survey. I looked again and saw only leaves and lichen-covered branches. Yet something had moved. Sure enough, there it was again.

A bird with brilliant scarlet plumage was darting from twig to twig in a small tree. It was a male Peruvian cock-of-the-rock, one of the most magnificent birds of South America. A more common species, the orange cock-of-the-rock, is found in Venezuela and Guyana, but the scarlet variety is limited to very localized patches of the subtropical Andean forests. I was lucky to see one. The bird remained low in the tree for some minutes and then, almost as if it had fallen from one of the branches, it vanished downstream.

We had to climb a steep slope in order to avoid a noisy cataract. Pushing through a barrier of blackberries, I noticed that some of the

stems had peculiar orange growths on them. When I looked closely, I realized they were insect galls. Insect larvae had hatched under the bark and emitted secretions that stimulated the cells of the host plant to grow into these unpleasant-looking abnormalities. I had seen other brambles in the cloud forest, but it seemed that only this one was affected by the insect galls.

INSECT GALL

Here, a pair of hummingbirds had been attracted by a fuchsia bush. Often these tiny birds peer inquisitively, hovering at eye level, but this pair were clearly not interested in us. For a moment, they hovered close to the fuchsia, their bodies suspended on a blur of rapid wing beats, and then were gone, leaving the long, pendant flowers unvisited.

From my vantage point in the middle of the stream, I looked back up the valley. On each side, the dim outlines of near vertical slopes towered through the mist. There was no going back, and the only way forward was to follow the stream through another canyon. A sudden red flash at the riverside halted my

MALE PERUVIAN COCK-OF-THE-ROCK

BACTERIAL GROWTH ON MOSS

We regained the comparative freedom of the rock-filled river, and carried on into the canyon. Within a few yards it had narrowed to ten feet. The wet, grey walls hemmed us in tightly, and I noticed a line of flood debris at least ten to 15 feet above us, marking the level of the torrent during the summer rains.

I followed Peter and Marion as they clambered down a disorderly pile of huge boulders. The stream rushed through the jumbled labyrinth beneath the rocks, producing a roar that reverberated deafeningly against the sheer sides of the canyon. Instinctively we glanced up at the flood level mark and hurried on.

After half an hour, to our relief, the valley opened slightly and another, powerful stream came in from the right. Quickly we realized that the force of the water was becoming too great for us to continue safely along the Challana. All the pools were deeper now and the waterfalls crashed down from ever increasing heights.

On every side, water dripped, poured and rushed into the valley. One heavy trickle spread across a rock, producing a blood-red stain. The rock was covered with a spongy, fibrous mass—probably a moss—and when I prodded it, a rich, red pigment ran out with the water. At first I thought the colour came from a deposit of some haematitic mineral, but I concluded that a growth of red bacteria was the cause. Some bacteria thrive on acidic water, and the run-off from the decaying, peaty mountainside would be acidic.

Stranded in the Cloud Forest

Our expedition was becoming extremely arduous. We had descended to about 5,000 feet, covering a distance of some nine miles, but already it was 4 o'clock. There was no hope now of reaching the bottom of the cloud forest before nightfall, so we prepared to camp.

We had to find trees from which to sling our covered hammocks, but the only suitable ones seemed to be tall, slender Cecropias whose hollow stems are often inhabited by stinging ants. These stinging ants are advantageous to the tree, protecting it from the ravages of leaf-cutting ants which work in long processions and can rapidly strip a plant bare. We saw none of these, but we noticed that some other insect scavengers had methodically eaten around the stems of some dead Cecropia leaves. As we did not relish the

CECROPIA LEAF RAVAGED BY ANTS

thought of a disturbed night's sleep, we spent nearly an hour looking for an alternative campsite, and eventually chose an area on a rocky shelf ten feet above the water.

With the approach of evening, the cloud forest became sullen and oppressive, and I think we all found it hard to hide our apprehension. We were, after all, alone on a remote mountainside where nobody ever came. We were still 5,000 feet above the Amazon lowlands, and nearly

A RED-TINGED BROMELIAD

NERTERA GRANADENSIS BERRIES

was *Nertera granadensis*, a relative of the European madder. It belongs to the Rubiaceae family, represented most widely in South America by Cinchona, the quinine bark, or Jesuit's bark as it was known when the Spanish first took it back with them to Europe.

About 150 yards downstream from our camp, the river plunged into a deep cleft, leaping over a wide lip of rock and forming a waterfall that cascaded for at least a hundred feet without interruption.

Once we had slung our hammocks and cleared some of the undergrowth beneath them, we set off in the fad-

twice that distance from our starting point. The wet season was about to start, and an isolated storm anywhere in the mountains above would quickly cut off our line of retreat. Once the rains established themselves, the forest would become impassable and our campsite would be cut off from above and below.

We felt imprisoned by plants. They even grew on the smoothest stone. Some mosses were no more than layers of green velvet. Many ferns had fronds blotched brown by their reproductive organs, from which would come a dust of millions of spores. Bromeliads with their trough-like rosettes of leaves clung to trees and cliffs and, where I was able to reach them, I looked into the water they had trapped at the base of their leaves. Small insects and forest debris floated there, providing nutrients for the plants. One low bush with bright red, glassy berries,

THE FOREST-SHROUDED WATERFALL

UNIDENTIFIED PLANT

branches, with the toes down.

Oropendolas belong to the New World Oriole family and are related to the grackles and cowbirds of North America. Finding them here was a sure sign that we were getting close to the lowland rain forest, where they are more common.

For a few brief moments the cloud thinned and left patches of faint blue sky. From where we stood at the edge of the waterfall, we could look up and see the snowy summit of a silent *nevado* thousands of feet above us. Then a heavy layer of cloud moved across the sky and the mountain

peak vanished once more.

While Marion and Peter went back to prepare supper, I explored our surroundings in more detail. Peering over the edge of the waterfall, I could see that the Challana plunged into a deep ravine running almost at right angles to the valley we had descended. On either side of the waterfall a steep cliff extended as far as I could see, and from my vantage point I looked down on to the tops of trees. Smoky traces of cloud drifted over the unbroken canopy below and hung in loose pockets on the opposite hillside.

ing light to explore the area around the waterfall. Where a soft mossy slope was drenched by the constant fine mist from the falling water, Marion collected several small, round nuts; half a dozen white fungi, rather like miniature puff balls; and a hard, ovoid vegetable mass that was something of a mystery. I believe it may have been a fungus. We took the specimen with us, but lost it later in the journey, and none of the experts could identify it from Marion's photograph.

High above the cascading water, the hanging nests of some oropendolas swung precariously from a tree and, as we approached, two of these yellow and black birds glided away into the forest below, their clear, fluty calls echoing lightly around the valley. The unusual nests are woven by female oropendolas from leaf fibres, and when they are finished, they look not unlike socks hanging from the tree

NESTING OROPENDOLAS

Somewhere beyond, the Challana flowed steeply eastwards to join the river Mapiri. Eventually this would leave the Andes and flow into the giant river Beni, and finally into the Amazon. I could see the stream running away at the bottom of the ravine, and I wondered idly how long it would take to reach the Atlantic, more than 3,000 miles away. Many, many months, I assumed.

It was clear that we could go no farther downriver without ropes and climbing equipment. There was no alternative for us but to climb back up the mountainside. Instead of following the tortuous stream bed, though, I decided we would hack our way a short distance up the side of the valley until we found the old mule trail again. This should not be too difficult and by following the relatively easy trail, we should be able to reach the top of the mountain within two days.

It was now nearly dark. As I left the waterfall, I had to wade through the stream again, but it had lost its mountain chill and seemed almost tepid. After the dry, rarefied air of the early morning *sierra*, the atmosphere down here was balmy and breathing was easy again. I sloshed forward quickly, and it did not seem to matter that the water was well above my knees. A breath of wood smoke and coffee drifted through the trees and there was a bright, luminous flash from a firefly close to the ground. I relaxed for the first time since dawn, thankful for the opportunity of a rest before starting the ascent the next day.

THE EDGE OF THE RAVINE

6/ The Hostile Plateaux

From midday onward through the stifling afternoon, the desert flings at your eyes a blaze of hard, yellow light reflected from its shining, salty crust.   T. HARPER GOODSPEED/ *PLANT HUNTERS IN THE ANDES*

It was a late afternoon in June, the Andean winter. I sat among clumps of spiky *ichú* grass, high on the stony slopes of Mount Illampu in the Cordillera Real, and gazed out over a grand sweep of Bolivian countryside. Far below me was the Altiplano, the great inland plateau which lies 12,000 feet up in the central Andes. In the 500 miles between here and its southern end, the Altiplano becomes progressively drier, ranging from semi-desert country to a bleak wilderness of giant salt flats or *salares* on which nothing grows. The following day I was to begin a journey to the greatest of these salt flats, the Salar de Uyuni, which covers more than 3,600 square miles; but for the moment I was quite content just to sit and look.

Fifty miles away on the opposite side of the Altiplano rose the Western Cordillera, a range of active and dormant volcanoes that bars the way to the Pacific. Their snowy summits glistened pinkly in the setting sun. The sky was as clear as a polished lens and from my eyrie I felt I could see the whole world. A sudden metallic flash far to the south-west, on my left, was the reflection of sunlight from the wings of an aircraft leaving El Alto, the airport of La Paz. The aeroplane was at least 60 miles away and not a sound from the engines interrupted the silence.

The Altiplano here is at its narrowest—hemmed in by the two snow-capped ranges—and it is largely filled by the silvery Lake Titicaca, 110 miles long by 35 miles wide. Titicaca has the distinction of being the highest navigable lake in the world, a title that dates back to 1872 when the first steamer to enter ferry service was carried up the mountains on mules, section by section, and reassembled on the lake shore. To me, Titicaca is the pivot of the Andes, a landmark for travellers and a natural staging point for Andean journeys. Titicaca is central in other ways too. The Aymara Indians who farm its harsh shores believe that their white-bearded god Viracocha rose from the water's chilly depths to establish their culture. The lake is an oasis amid the surrounding deserts, a haven for the many water birds that seek refuge among the dense *totora* reeds of the sheltered bays. It even has a species all to itself, the Flightless Titicaca Grebe. Near Puno on the north-west side of the lake, I once caught sight of this bird battling towards the cover of the reeds. All it could manage in its haste was a laborious paddle flight that never quite lifted it from the water.

Titicaca is sustained by the normally dependable summer rains and by numerous icy mountain streams. In one recent year, however, the vital balance of climatic conditions was upset. The summer was exceptionally dry and also rather cold, with the result that the mountain snows failed to melt as usual. Then came the keen Andean winter with its clear skies and dry air. When there is no screen of protective rain clouds, moisture evaporates very quickly on the Altiplano. At such a high altitude, the atmospheric pressure is low and the sun's rays are especially intense, and the air acts like a sponge. By the end of the winter, lack of rain and rapid evaporation had caused the level of Lake Titicaca to drop four feet—the equivalent of 11 thousand million tons of water.

The drying process, so dramatically illustrated by that exceptional year, has been continuing inexorably for millennia. Titicaca, impressive though it is, was once part of a far larger body of water. About 15,000 years ago, towards the end of the last Ice Age, two inland seas filled the Altiplano to a level over 200 feet above the present floor of the basin. Lake Ballivian spread across the north, including the area of Titicaca, and Lake Minchin reached from the central Altiplano down to what are now the plains of southern Bolivia. These seas were fed by the huge volumes of water which flowed down from giant snowfields and glaciers during the late Pleistocene summers. But after the Ice Age, such liberal replenishment stopped and gradually the sea-lakes dried out.

The Altiplano of today contains very little water apart from Titicaca and a scatter of fresh-water lagoons, the last remnants of Lake Ballivian. In the south, nothing is left of Lake Minchin except Lake Poopó, a series of dried-out mudflats, and the giant salt flats such as

Uyuni. Except for two to three months in the summer, when there may be heavy storms, rain or snow virtually never falls in the Altiplano. No moisture comes in from the Pacific and the rains from the Atlantic are caught by the eastern cordilleras. For most of the year, the intense sun and thin mountain air siphon off the last drops of humidity. When the sun goes down, the heat of the day disappears rapidly and the nights are bitterly cold, well below freezing.

Beyond the Cordillera de Lipez, the southern barrier of the Altiplano, lies another region of inland drainage, the Puna de Atacama. This *puna*, or wild open place, lies even higher than the Altiplano, at 14,000 feet. It is a desert composed of remote plateaux separated by volcanoes. Seasonal streams flow into it from the surrounding mountains but no rivers flow out, and the water is therefore trapped, either collecting in lakes and marshes or more often evaporating to form countless salt lakes. The most fascinating of these is the Laguna Colorada, or Red Lake, which lies in the extreme north-west of the Puna de Atacama.

During my journey south from Titicaca, I hoped to see much of this wild mountain country, especially the Salar de Uyuni and the Laguna Colorada, but the trip would not be easy. Large areas of the southern Altiplano and the Puna de Atacama are uninhabited and some parts are little explored. For most of the way there would be no roads, and even the best maps are unreliable for this section of the Andes. At such heights the winter cold would be extreme, enough to warrant arctic sleeping bags, and we would have to go without washing for several weeks. Water evaporates so quickly that it creates a refrigeration effect, and moments after a wash in a hot spring, your hands are as cold as if you had plunged them into icy water. In addition, I was doubtful about sources of drinking water. On previous trips I had seen enough evidence to persuade me that the Altiplano is still drying out. In the great days of Spanish America, when this was silver-mine country, there were villages with good wells, but today the stone-built houses of the miners stand abandoned in waterless valleys.

I had been sitting on Mount Illampu for a long time and the ground was hard. By now the sun was suspended just above the Western Cordillera, and within a few minutes it had dropped below the peaks. The air, already cold, chilled a few degrees more. The change was instantly perceptible, and I wondered how much colder it would be by midnight.

The following day Marion and I packed the Land Rover with a month's precautionary supplies: water, food and 40 gallons of extra fuel. We set out from La Paz, hoping to reach the salt flat of Uyuni in three days of

The saline Laguna Colorada, with clumps of ichú grass and a freshwater pool (foreground), lies in an arid setting typical of the high plateaux.

almost non-stop driving. The first day out, we used the tarmac road that runs for 150 miles to Oruro, the only town on our route, and then stops. We passed from the relatively benign countryside of Lake Titicaca into a barren world of wide open spaces. To the west we saw the great volcano Sajama, its snow-covered summit towering to 21,390 feet. As we continued to the south-west, we were able to see parts of the sluggish river Desaguadero, the only outlet from Titicaca. It flows from the southern end of the lake for 200 miles before depositing its water into Lake Uru-Uru, a new body of water just north of Lake Poopó.

When we reached Oruro, I felt we had come to the farthest edge of civilization. The hills around the town are streaked yellow and dark brown with the spoil from the local tin mines, and as we entered the outskirts we passed lorries crowded with miners. Ahead of us, there would be nothing but ghost towns and a few isolated hamlets. We crossed the river Desaguadero by the bridge outside Oruro, and drove west along a dusty desert track that skirts Lake Poopó, now largely dry. By evening we expected to reach the river Lacajahuira. Most maps show it running between Lake Poopó and the Salar de Coipasa, further west. To our surprise there was no river. It too was dry.

On the third day we came to the Salar de Uyuni. We approached it from the north, skirting the cone of Tunupa, an extinct volcano more than 17,000 feet high. As we climbed a rocky shoulder, we caught our first sight of the salt flat. The rocky foreshore sloped down to it, much like the shore of any ordinary lake, but where water would have lapped against stone, there was a flat surface shimmering white and stretching as far as the eye could see.

The Salar de Uyuni is not so much a salt flat as a sea of salt, 85 miles across at its widest and about 100 miles long. It is the largest expanse of salt in South America, and its size is not surprising when you consider its origins. After the drying out of Lake Minchin, which flooded this part of the Altiplano right across from the eastern cordilleras to the Western, a giant expanse of flat lake-bed was exposed to the sun and the rarefied air. For more than 10,000 years, intense evaporation sucked the surface dry, drew up more moisture from below, and siphoned that away too. In the process a strong concentration of minerals was brought to the surface. These minerals consisted mostly of gypsum and common salt that had soaked into the ground from the surrounding volcanic cordilleras after rain had fallen. When they met the air, they dried out and crystallized, and the salt, the last to crystallize, was left at the top.

The process continues even now, and it is dramatically demonstrated when the rains are heavy. On such occasions the *salar* is covered with several inches of water, but it is so effective as an evaporation pan that thousands of tons of water are drawn off every day until the surface is left coated with a new film of salt brought up in solution from below.

After so many thousands of years, the salt crust is mostly between six and 24 feet thick. Beneath it lies a deep basin of sedimentary rock, gravel and mud, several thousand feet deep in places, which holds water like a sponge. Near the shore, the salt crust becomes much thinner, forming a fragile covering over treacherous, oozy mud. Our plan was to drive across the Salar de Uyuni from north to south, examining its surface as we went and speeding up our journey into the Puna de Atacama. But first we had to find a suitable point to enter the lake. Judging from the map, the most direct route was through the small *pueblo* of Salinas de Garci Mendoza, but at this village there would be no way of crossing the mud zone. Instead we would have to find one of the causeways built at strategic points by the occasional lorry drivers who take short cuts across the *salar*. I had been told there was a causeway near a tiny Indian settlement directly south of Tunupa. Sure enough, as we breasted the last rise before the *pueblito*, we saw a line of stones running out on to the salt.

We drove down the causeway, and stopped on the *salar* so that I could walk back and inspect the mud zone at its edge. I dug into the dry salt with my heel and broke through. Underneath, the mud was black, soft and wet. When I took another foolhardy step forward, the surface crumbled and I went through up to my shins. It was a ready-made grave for the unwary. The cold of the mountain winter began to bite, and I was soon shivering. The temperature must have been near freezing point.

Back in the Land Rover I assumed the role of navigator while Marion took the wheel. We maintained a steady 50 miles an hour. At first I felt we were at sea, racing across a calm surface with the shore of the *salar* receding in the background. Then it seemed as though we were flying in a clear sky. The Land Rover's engine hummed as steadily as that of a cruising aircraft. Marion kept her foot pressed gently on the accelerator and her hands lightly on the steering wheel. The surface of the *salar* was so flat and even that she once or twice drove with her eyes closed, just to show that it could be done.

We had been travelling south-west for 45 minutes and the crests of the volcanic peaks along the Bolivian-Chile frontier were beginning to show as large black blobs above the horizon. After another 20 minutes, one of

A crimson Hoffmanseggia flower blooms vividly on the gravelly shore of an island in the salt flat of Uyuni. The tiny plant, a member of the pea family, lies low on the ground to escape the harsh wind and uses deep roots to absorb what moisture is available. The brilliant colour and strong scent of the flower attracts pollinating insects.

these blobs appeared to solidify. It seemed to be much closer than the rest, and I wondered whether it might be an island in the salt. Marion steered the Land Rover a few more degrees to the west and aimed for it. Half a mile away, I could see that it was indeed an island, about 150 feet high and a mile or so long, with tall columnar cacti on its hump. As we drove nearer I could see cacti on every slope. I felt like a sailor looking at a new-found land, and began to scan the island for signs of life. There was a salt bay almost directly ahead and I suggested heading into it. This was a mistake. The Land Rover slowed down sickeningly and began to sink into soft salt more than 400 yards from the shore.

Marion's reactions were quick. Straight down into second gear, low-range, four-wheel drive and keep the vehicle moving. Not forward into the morass but making a gently curving arc to reach the hard salt again. The wheels were gripping well. Suddenly we hit a very soft patch. The wheels spun. A fast reverse. More spin, then forward. Then stop before sinking in too deeply. Marion switched off the engine, cursed my navigation and pointed to the shovel, as if to say, dig us out.

Although it was late afternoon, the sky was still bright and there was bound to be good light for an hour or two. While Marion trekked shorewards with our camping kit, I set about securing the Land Rover to prevent it from sinking deeper into the mud underneath the salt. We were well equipped to cope with average "boggings" and carried an assortment of jacks, shovels, crow-bars, sledge hammers and specially strengthened metal channels. So much, in fact, that together with the 40-gallon drum of fuel, we had little room for food.

The first job was to jack up the Land Rover's rear wheels. Luckily I had one of those monster ratchet jacks used for lifting the leviathans of the Detroit automobile plants. It was simple, almost unbreakable, and gave an effective lift of nearly three feet. I placed it on a flat metal plate I carried for such emergencies and raised the nearside rear wheel clear of the salt. I laid one of our two spare wheels horizontally in the muddy patch underneath so that it would spread the weight, and let the rear wheel down. Instead of sinking into the mud, it came to rest on the improvised foundation. I successfully repeated the operation on the offside.

The effort had left me gasping in the rarefied air, and I stopped for a rest before grappling with the front end, which was more deeply embedded. Marion had meanwhile returned from the island, where she had left the camping gear. The distance was enough to make me abandon my first idea of building a makeshift causeway. To bring the necessary rocks would be too exhausting and time consuming. I considered the other possibilities. We could start exploring the island and try to arrange a rescue party. Smoke signals and flags would be useless in so remote a place, but we might possibly use a mirror to flash one of the commercial aircraft that fly over the *salar* from La Paz to Buenos Aires.

I went back to work and managed to jack up the front of the vehicle just enough to slide two metal channels under the wheels. I placed two more channels behind the rear wheels and asked Marion to reverse. The plan worked. She drove easily and quickly out of the morass, and our troubles came to an end as suddenly as they had begun.

I walked part of the way around the island, prodding the salt with a crowbar to find a firm surface, and Marion drove slowly behind. We found that the salt was thickest near a rocky promontory and we were able to stop the Land Rover within a yard of the shore.

The fine grey sand of the beach was scattered with a few plants and yellowing tufts of *ichú* grass. One small crimson flower peeped out from below the sandy surface. Above us the rocky slope was covered by scraggly *tola* bushes and cacti, which I recognized as a type known locally as *cardón*. I left Marion to set up camp on the shore and clambered up to investigate a ledge whose light colour had attracted my attention. The rock was covered with a greyish-white encrustation. To examine this material more closely, I took a large stone and struck it against an encrusted boulder which stood out from the rock face. A hard outer concretion about two inches thick fell away, exposing a dark grey inner core of rock underneath. I concluded that the crust was the calcareous remains of a prolific growth of algae. Long ago, waves had splashed against the cliffs where I was standing. The algae had established themselves both in this splash zone and just below the surface where sunlight had penetrated, permanently marking the level of the vanished inland sea, Lake Minchin.

Looking around, I noticed that the calcareous encrustation stopped only a few feet above me, forming a band right along the side of the island. Beyond this line the rock was bare. I scrambled up the rocky slope, slipping on loose stones, to find out whether there were any calcareous marks higher up. By the time I reached the top of the island I had found two more distinct bands. Relatively little is known about the geology of this area but it seemed clear that each of these bands represented a period when the level of the ancient sea was stable and higher and the water was warm enough for algae to grow. Perhaps there was a temporary improvement in the ancient climate, when heavy rains and meltwater from the shrinking glaciers kept the level of the lake steady. Then came a severe drought lasting hundreds or even thousands of years, a period of constant evaporation when the algae had no chance to flourish at any one level. This might well explain the wide stretches of bare rock between the whitish bands.

At the top of the island I was surrounded by the tall, sharp-spined cacti I had seen from the Land Rover, and by a multitude of *tola* bushes and tough spiny plants. Beyond lay the dazzling ocean of the *salar*, stretching flat in every direction except for the flecks of a few other islets.

As I scrambled down the steep slope again, I discovered a small rock shelter just above the highest of the calcareous bands. On the ground were scattered some broken flamingo eggshells. Then I found some perfect arrowheads. The jagged edges of the tiny blades were as clean and sharp as when they had been chipped. They were the tools of some early Andean hunter who must have come here in a simple boat in the last days of Lake Minchin, or else visited the island once the waters had receded. The relics had lain undisturbed in this spot ever since the huntsman dropped them.

The next morning we left the island, bearing due south across the *salar*. Occasionally we stopped to look at unexpected features of the surface. In several spots near the eastern edge of the *salar* we found shallow pools of clear water, fed perhaps by streams flowing underneath the salt. Some were filled with wonderfully large salt crystals

The compact shape and thick skins of barrel cacti protect them from extreme heat and cold. In addition, some have concertina-like stems that can expand to store water. The spines serve as a deterrent to any grazing animals.

A SPECIES OF MATUCANA

MELOCACTUS TRUJILLOENSIS

TRICHOCEREUS BERTRAMIANUS

more than half an inch across and shaped like stepped pyramids. At one pool I lifted two or three of the crystals out of the water. For a moment they sparkled, then lost their lustre in the dry air, leaving a film of salt on my hand. As soon as I put them back in the water, their brilliance returned. When we stopped another time, we found fissures in the salt filled with cold, deep blue water. I could see the knobbly salt walls of the fissures going down some feet below the surface, and wondered how deep the cracks were. I plumbed one of them with a spanner tied to 20 feet of nylon cord, but could not reach the bottom. Not far away I came across water flowing from one fissure, welling gently up and spreading over the surface of the *salar*. I tasted it and found it less brackish than I had expected. Considering that it was surrounded by so much salt, it was surprisingly fresh.

We made good time across the remainder of the *salar* and by mid-morning had reached its southern end. We drove on to dry land across another rough stone causeway and set out on the next leg of our journey. Our goal was the Laguna Colorada, but three or four days of difficult travel would be needed before we reached it. We had to cross the remainder of the Altiplano, climb into the high Puna de Atacama and then drive for a hundred miles almost to the Western Cordillera, at whose base the lake is situated. Our first task was to negotiate a huge mudflat, the Pampa Colorada, and then ford the Río Grande de Lipez, the only river of any size to flow in the southernmost Altiplano.

A Darwin's rhea paces across an arid plateau 14,000 feet up in the Altiplano. In such open spaces, speed is the main protection of this flightless bird, which resembles an ostrich and could out-run a horse. At nesting time, each male acquires up to six females which lay their eggs in the same hollow. The male then incubates the eggs until they hatch.

During the summer rains the Pampa Colorada turns into a treacherous red paste and becomes impassable. With the onset of winter, when the water is evaporated away, the mud returns to its more usual hard, dry and cracked state, and this is how we hoped to find it. For the first 15 miles we drove fast along a rough track made by trucks travelling to and from a mine near the Bolivia-Argentina frontier. The trail skirted many rocky outcrops that reared out of the mud, but in one place it crossed a low hill about 2,000 feet above the Altiplano. We stopped the Land Rover on this vantage point and I looked back for my last glimpse of the *salar*. It lay behind us, a thin white bar lodged between the mountains. At its far end I could still see the old volcano of Tunupa. To the west were some low hills, the Cordillera de las Minas, and beyond, the peaks of the Western Cordillera itself, a line of dark volcanoes, some exhaling wisps of steam, others quiet and brooding. To the east the horizon was flat. In the south were the mountains that marked the far end of the Altiplano—the Cordillera de Lipez, dominated by the summit of Nuevo Mundo that stands at over 19,000 feet high.

We drove on across the mudflat at a comfortable 50 miles an hour. Suddenly the surface ahead of us degenerated into a series of rough waves. The mud had been churned like margarine into trenches and mounds by a large truck. It was still soft, and we would have to make a detour. We turned south-east towards a flat horizon. The hills which had been our reference points were soon no more than ill-defined outlines shimmering in the distant haze. Luckily I had taken a bearing from the point of our diversion and from a large-scale map I worked out a new course. If we turned through something more than a 90-degree angle after passing the area of soft mud, I estimated that we should meet a curve of the Río Grande de Lipez within 20 miles.

We had some nasty moments crossing shallow gullies with ominous dark red mud at the bottom, but we did not get stuck. We travelled more slowly now, so as to conserve fuel, but in an hour we reached the end of the *pampa*. Then we bumped and jolted for another hour through soft sand and thick *tola* bushes. As if to compensate for the detour, we surprised a group of rheas which rose from their camouflaged resting places and hurried off like clockwork feather-dusters. The rhea of the Altiplano, a rare, high-altitude race of Darwin's rhea, resembles a shrunken African ostrich and stands only four feet high. It relies on its superb mottled grey colouring to camouflage it from its enemies, which include foxes and mountain cats. When pursued, rheas can run at a demonically fast speed. They have a curious but effective habit of doubling back on their escape route and then lying flat on the ground, merging with the *tola* bushes. In the soft sand we had no hope of getting closer to them, and we continued on our compass bearing until we joined up once again with the sandy mining trail we had left so many miles before.

We arrived at the river and braked on its dry, sandy bank. It was about 50 yards wide but I had no idea how deep it was, so I rolled up my trousers and waded in to get an idea. The water was agonizingly cold. Chunks of ice floated past me as I moved gingerly forward, and within seconds my feet were numb. Fortunately the river turned out to be only a few inches deep even in the middle and I was able to splash back to the Land Rover without suffering an even more marrow-chilling immersion. With a string of sharp reports as the vehicle smashed through the thin ice at the banks, we drove rapidly across the Río Grande de Lipez.

Before leaving the Altiplano, we briefly visited a small colonial Spanish mining town. More than two centuries ago it was a flourishing community, rich on silver from the near-by mountains; now it is a tumbledown village inhabited only by a few Indians and a Polish field geologist, Piotr Zubrzycki, whom I had met in La Paz. Piotr was one of those rare individuals who possess a sensitive contact with the landscape. He had been surveying the high deserts for more than 15 years, covering most of his territory on foot, and was clearly at home in the cold open spaces. A short beard protected his face from the biting winds and strong sunlight, and the exposed skin was brown and dry. As we walked down an empty street, between the crumbling walls of houses long since abandoned, he offered to take me to one of his special haunts, the home of an old silver prospector who had died 30 or 40 years ago. The house had not been touched since. Only Piotr had access to it and the Indians kept away out of superstition. It was in another, even more derelict village not far away.

We set off that afternoon, Piotr walking with effortless strides. When we reached the village, it was late evening, dry and very cold. Piotr knew his way, even at night. We came to a small adobe house with a locked door. He fumbled through the pockets of his windproof jacket for a key, found it and smoothly turned the lock. The door swung silently open and we walked into a dark, quiet room. Piotr found some matches and lit a candle. Around the walls were shelves tightly filled with books in English, Spanish and Polish. There were newspapers, brown and brittle. In a corner, card boxes were piled, each one filled with specimens and rock samples. An ancient phonograph and a case of records sat in another corner. Near by on an enamelled washstand lay a cracked cake of soap and an open razor, its blade still untarnished.

Piotr pointed to two photographs on the wall: an old man and his wife. "They came from my country," he said sadly, "and this is where the prospector wandered for most of his life." I stared at the piled notebooks and rolls of hand-drawn maps. Whatever he had come to search for, the old man had not found it. But I imagine that he did discover spiritual tranquillity in this lonely part of the Andes. He seemed to have left some of it behind in that room. Some intangible yet powerful part of him seemed to hover there—a definite presence—I sensed it without a word from Piotr. The room was quiet and still. We turned and stepped back into the sharp, clear-mountain night, leaving the tiny house to the wilderness.

Before we left the deserted mining village, Piotr asked me where we were going. I outlined our plan of driving up into the Puna de Atacama to see the Laguna Colorada. "A good idea," he commented. "You might see the rare little flamingoes, the James's, I believe. Besides, the lake itself is an astonishing sight." He laughed. "But cold. Very cold. Don't forget

The compact surface of the llareta plant, dotted with miniature yellow flowers, smooths the sharp corners of a desert boulder. Related to parsley, the slow-growing llareta has miniscule leaves that minimize the amount of water loss in the arid areas it favours.

your *pisco*." *Pisco* is a fiery local spirit, and we had a good supply.

Cold or not, I was interested in getting to the lake, for although I had seen the commoner Andean and Chilean flamingoes on previous occasions, I had never had any luck with the James's. We drove south next morning along the lazy Río Quetena, a tributary of the Río Grande de Lipez, and then climbed from the Altiplano to an open plateau 15,000 feet above sea level. This was the Planicie del Panizo, which extends for several hundred square miles and is the northernmost region of the Puna de Atacama. Occasionally I caught glimpses of golden-fleeced vicuña dashing for cover between the rocks, creating miniature explosions of dust beneath their hoofs, but that was all. For most of the time a vast barren desert stretched grimly before us.

Around noon we took a short rest. I stood and watched a bank of clouds in the eastern hills with some curiosity. In winter the sky is generally clear in the *puna*, but I had no time to think over the anomaly of these clouds. Within seconds, a hail of ice particles, sand and salt blasted us back to the Land Rover. As we ran, we held our anoraks over our faces in an attempt to shield them from the stinging sharp snow. We had been hit by the deadly *viento blanco*, or "white wind", a notorious feature of the high *puna*. For half an hour we were besieged in the Land Rover, which sounded as though it was being sandblasted.

When the scouring storm had passed I noticed a heap of rocks that had a peculiar, bright green, moss-like growth on one side. I recognized it as a *llareta* plant. Its surface was composed of thousands of tiny fleshy leaves growing in minute rosettes. Oozing out between the rosettes were globules of clear resin. The *llareta* was speckled with groups of miniature yellow flowers, each about two millimetres across. With a sheath knife I tried to prise a small piece of the plant from the rock, but it was so tough that the blade made hardly any impression. I returned to the Land Rover and collected a hammer and tyre lever which I used to better effect, digging away a tiny chunk. Under the green surface of the *llareta* was a hard, brown, fibrous mass—accumulations of many years' growth, one on top of the other. This material is so wood-like and resinous that it makes an excellent fuel.

Beyond the *llareta* plant, the surface of the plateau appeared flat, but I walked on a little in case anything else interesting should turn up. The ground was covered with small pieces of ignimbrite, like giant corn-flakes, and with every step I crunched them underfoot. By now all traces of the *viento blanco* had evaporated in the arid air and sandy dust was swirling again. But I soon came to a slight dip where drifts of snow had

The bowed shapes of so-called penitent snow in the Puna de Atacama remain long after the last snowfalls of winter. The shapes are carved by cold wind and sun which gradually deepen small depressions on the snow's surface until the pillars are completely free-standing.

collected in the summer. Instead of forming a smooth carpet, however, the snow had turned into a field of dusty stalagmites, each about a foot high and gently bowed. These little pinnacles are known locally as penitent snow because of their resemblance to nuns at prayer, and they are formed through sublimation, a rapid evaporation process in which ice turns straight into vapour without melting. When a snow-bed is subjected to the sun's rays, small depressions are formed. If there is no new fall of snow, the depressions deepen and their hollows merge, leaving towers of ice between them. As the sun continues to warm the base of the depressions, the ground is reached and the penitents begin to stand out as individual structures, ever being sharpened by the evaporating effect of the powerful winds. Because it is evaporation that shapes the snow, the ground beneath each spire was dry. I was later told that the penitents can reach heights of up to 20 feet.

After our rest, we turned due west and in two days we were approaching the volcanic Western Cordillera. The Laguna Colorada lies among their gentle foothills, and soon we caught sight of it, resting in a shallow depression about ten miles ahead. The lake was bright red and as we came closer, bumping and sometimes almost falling, down a staircase of ignimbrite sheets, it began to look more and more like thick tomato soup. When we reached the steep shore, thick creamy waves were lapping on the beach. I put a sample of water into a bottle and held it up to see whether the colour was an illusion. Against the light it shone a distinct pink, sufficient to account for its redness in the lake, where the depth of water and the brown of the surrounding hills make it appear darker.

The secret of the Laguna Colorada lies in its salts, which nourish a massive concentration of reddish algae. From a higher shelf on the sharp slope I could see great blocks of salt-coated gypsum hundreds of yards long standing on the far side of the lake and reaching into its centre like stranded icebergs. These floes, shimmering a hallucinatory white against the red of the lake, must have been formed by rapid evaporation during some ancient drought. Through my binoculars, I could see even more salt covering a low peninsula that partly divided the lake farther south. The place was a gigantic cauldron from which the cold dry air had sucked moisture as fast as some hellish fire would have boiled it.

From my vantage point I could see most of the lake's perimeter. There were no signs of shore life apart from a mat of flamingo feathers littering the sand below me, and I began to think that perhaps the rare James's flamingoes had been driven away by the extreme cold. I scanned the lake

A flight of James's flamingoes passes low over the vivid surface of the Laguna Colorada, a major nesting ground of this rare species. Innumerable pink algae give the salt water of the lake its characteristic colour and provide the main food of these birds.

Chilean and Andean flamingoes (the latter identifiable by their black-tipped wings) wade through a freshwater pool on the edge of the Laguna Colorada. The birds come here from time to time to drink and soak off the salt that builds up on their legs.

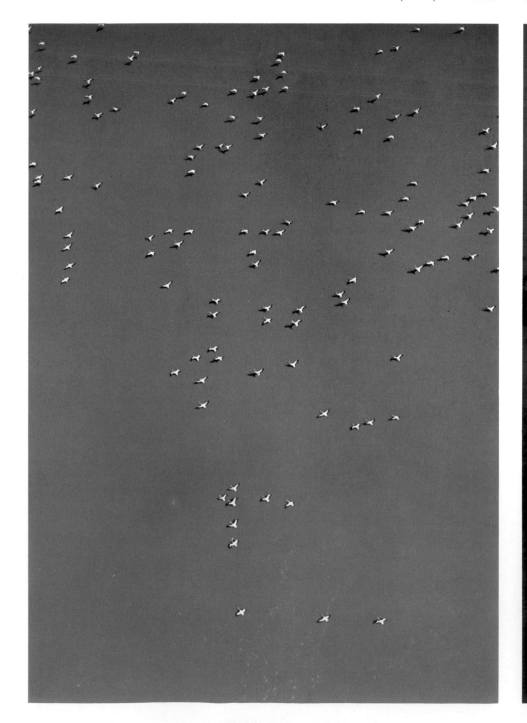

again with my binoculars, and this time spotted a group of about a hundred birds wading in shallow water. They were the common Chilean and Greater Andean flamingoes, quite tall with grey and yellow legs. I panned the binoculars again and found another group, the smaller James's flamingoes, beautiful birds barely three feet tall with brick-red legs, a distinctively large patch of bright yellow on the bill, and paler pink feathers. They were named in 1886 after Berkeley James, a British businessman who sponsored the expedition that discovered the species in the high Atacama. After that first sighting, the birds remained unrecorded until they were rediscovered in 1957. The mountain wind ruffled the birds' feathers as they swept their beaks through the icy red water, sifting the microscopic diatoms and other algae on which they feed. In such a savage environment it was not surprising, I reflected, that they had been missed for so long. Few people would want to come here.

We decided to spend the night by the lake and pitched our tent in the lee of a giant boulder, pinning the sides to the ground with a dozen heavy rocks. By afternoon the westerly wind had redoubled its force and we faced another terrifying icy blast that whirled sharp, grey sand through the air. We took shelter in the tent and peered through a chink in the entrance flaps, watching the brilliant hues of the lake turn into a palette of turbulent colours. A tall column of powdery salt crystals spiralled up from a distant beach and billowed into a fast moving cloud that swept across the water, changing it from red to pink and then grey. A hail of sand battered the canvas, and when we dared to look out again, all we could see were some clumps of yellow *ichú* grass a few feet away. Some of the weaker blades were bent under the force of the wind but the tougher spikes just quivered as the sand piled up rapidly on one side and was scoured away on the other.

These violent winds are a characteristic feature of the high inland plateaux. Because the rarefied mountain air cannot retain warmth, the temperature always drops sharply when the sun sets behind the Western Cordillera. First the east-facing slopes and then the floor of the *puna* are plunged into shadow. The result is a powerful katabatic, or downhill wind. As the air on the darkening slopes rapidly cools, it contracts and becomes denser and therefore heavier. Consequently it starts to flow downhill. Meanwhile, as the shadows lengthen, the warm daytime air rises away from the *puna* floor, creating a low-pressure zone that the cold air from the slopes rushes in to fill. Once the entire *puna* is in darkness and all the air has cooled, an equilibrium is established and the wind abruptly stops. No doubt that was why the canvas of our tent went suddenly limp. We crept out to find that the sun had passed behind the Western Cordillera. A deep red-violet glow suffused the sky, making a silky backdrop for the first stars of the Andean night.

In the morning the thermometer registered zero Fahrenheit, and we could hardly bear to peel ourselves from out heavily padded arctic sleeping bags. I had to thaw some water before we could make a pot of miner's tea—strong, with sugar, fresh lime juice and *pisco*. We drank it tepid because the high altitude (about 14,800 feet) had reduced the boiling point of water. It was so cold that when we washed our metal dishes, they acquired a film of ice before we could dry them.

From the Laguna Colorada, we ventured deeper and deeper into the Puna de Atacama, travelling for more than three weeks until we reached the 22,579-foot Ojos del Salado. It is the highest volcano in the world and marks the southern end of the *puna*, so it seemed a natural finish to our journey through the high deserts of the Andes. As we continued south, the intensely cold, dry air began to leave its mark. My hands and lips dried out and cracked in spite of repeated applications of high altitude creams: none seemed to work in the conditions for which it was recommended. Where my skin was inflamed by the sun, it tingled red and irritated strongly, finally turning a deep brown.

Our rarefied desert environment seemed unnaturally bright, with deep blue, almost purple skies against which the land appeared pale, sometimes nearly white. There was no sign of rain or snowfall and many of the deep-cut river valleys were dry. Occasionally, though, we found damp places where water had seeped down from small snowfields or from tiny springs. Here the ground laid soft carpets of brilliant green on the desert. Often the rocks were a rusty ferruginous brown or a rich cuprous green, and the volcanoes we passed were smeared yellow with sulphur. Each day seemed to grow colder, dustier and drier. The nights were long, and outside our tent the world seemed to die. The great winds, so fiercesome during the day, left us each night with an uncanny stillness. The air was so clear that the stars never shimmered. We slept under the blackened, infinite roof of the Andes, utterly alone.

Fountains in the High Desert

The highest geyser field in the world, and the largest collection of geysers in the Andes, lies in the Tatio plateau, a small but unique expanse of desert 14,300 feet above sea level in the Western Cordillera of Chile. The desert's surface is punctured with vents from which jets of steam and boiling water spurt ten to 15 feet into the air.

The geysers are most impressive just after dawn, when the mountain air is still and bitterly cold. Columns of steam billow over the plateau, and in the silence of the early morning a steady hissing, plopping and gurgling is most clearly audible. Later in the day, the keen north-westerly winds blow more strongly, dispersing the steam and obliterating the sound of the water.

Tatio's geysers, unlike the better known types that erupt at predictable intervals, are constantly bubbling, and they have been active for tens of thousands of years. They derive their energy from deep underground. Water from snowfields on the near-by mountains and from summer rainstorms seeps through the porous rock of the plateau until it is trapped in a basin-shaped layer of impermeable rock.

At the bottom of this subter-ranean cauldron, the water comes into contact with intensely hot rocks which were once magmas feeding the neighbouring volcanoes. The water is under such enormous pressure that it is unable to expand and boil, and consequently it super-heats. Because hot water is lighter than cold, it begins to rise. When its pent-up pressure exceeds the weight of water above, it expands violently, rushing up through cracks and fis-sures to the surface. On its way, it dissolves silica and other mineral compounds from surrounding rocks.

The water bubbles out of the geysers at boiling point and at once begins to evaporate in the cold, dry wind. Gradually, the dissolved min-erals crystallize into tiny, bead-like concretions that collect around the vents, building up over the centuries into solid, chimney-pot structures.

The remaining hot water spreads out around the bases of the chimney pots, depositing more minerals which form a multitude of small terrace-like dams, each enclosing a pool of hot water. Farther down-stream the water is still warm and sustains a rich growth of algae that carpets the stream bed with deep browns and greens, creating small patches of luxuriance in the desert.

A column of steam, swept low along the ground by the mid-morning wind, hisses out of one of the Tatio geysers. The steam is mixed with boiling water, which at this altitude has a temperature of only 185 degrees Fahrenheit.

Around the double vent of a young
geyser (above), tiny beads of crystallized
silica and chlorides have built up into
a cone about 18 inches high. The beads
are produced as the strong, cold wind
evaporates the water in which the
minerals are suspended.

After hundreds of years, the constant
bubbling of mineral-laden water from
this old geyser has created a chimney
pot that stands more than four feet
high. Pinkish streaks on its rough
sides betray the presence of manganese
and heat-tolerant algae.

Delicate lace patterns have been formed on a shallow terrace close to the base of an old geyser. They are cross-sections of the original mineral encrustations which have been heavily eroded over many years by the hot water that is continually flushing out of the geyser.

About 200 yards from a geyser, the stream bed is carpeted with masses of algae. At this distance, the water is about 120 degrees Fahrenheit and the warmth, combined with dissolved mineral nutrients, provides an ideal environment for the tiny plants.

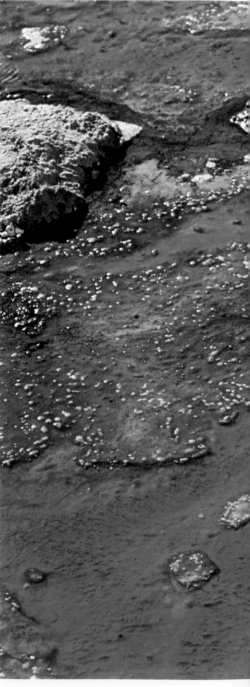

Morning sunlight casts a satin sheen on the rippled water that encircles one of the Tatio geysers. The water has drained down the slight incline of the plateau, merging with streams from other geysers; a large proportion sinks back into the porous ground where it may eventually be re-heated by hot rocks to spout up once more.

7/ The Uttermost Realm

These vast piles of snow, which never melt, and seem destined to last as long as the world holds together, present a noble and even sublime spectacle. CHARLES DARWIN/ *THE VOYAGE OF THE BEAGLE*

The earth tilted, hesitated and levelled out again as my small aircraft turned north. To starboard lay the dun-coloured Patagonian plain, stretching 200 miles to the Atlantic, its surface thinly patched with snow. To port was a range of granite peaks, slender and sharp as fangs, the Patagonian Andes: blasted smooth by wind, they stood sheathed in ice. Their slender pinnacles were crowned with crows' nests of frozen snow, and in the valleys between their flanks, glaciers flowed down; white tongues licking at grey rock. The sky was clear and I had an excellent view of Lake Argentino, a water-filled depression gouged at the base of the mountains by the Ice Ages. At the far end of the lake the Moreno Glacier reached down into the water—nine miles of ice disgorging a procession of icebergs that floated across the milky surface. Soon, I hoped, I would see this glacier at close quarters.

To north and south, the mountains hung motionless in the cold blue sky. We stayed at a respectful distance, mindful of turbulent airstreams which are often thrown off the bluff rock faces. The pioneer French aviator and writer, Antoine de St. Exupery, has described how, when he was opening up the South American air routes in the 1920s, the winds raging out of these mountains hurled his small aeroplane across the sky like a leaf in a storm. The thermometer mounted on the wing, he noticed, registered —20° F. No doubt the winter air was as cold today, but the winds were not so fierce and we continued steadily north.

From a map resting on my knee, I identified the mountains as we passed them: Cerro Bolados, Cerro Agasiz, Cerro Bertrand, Cerro Murallon and also Cerro Torre, exceptional for its needle-like form. Hidden just behind the mountains was the Patagonian ice-cap, its two sections forming the second largest expanse of ice outside the polar regions and one of the least explored wildernesses in the world. Beyond that, a mere 30 miles away, lay the Pacific.

We climbed as we approached Cerro FitzRoy, named after the captain of the *Beagle* in which the young Charles Darwin made his famous voyage to the southernmost Andes in 1832. The mountain's 11,000-foot summit was obscured by its characteristic wreath of cloud, a phenomenon that accounts for the ancient, though mistaken, Indian name of Chalten (Volcano). From its sheer face, slivers of rock had been stripped by glacial and freeze-thaw erosion and lay as rubble at the mountain's base. We had reached the furthest limit of our flight, and banked to begin the return journey.

This aerial reconnaissance provided my only overall view of the southernmost Andes. For about 500 miles, from latitude 46° south to the tip of the continent, these mountains are permanently icebound, and for weeks at a time they are ravaged by storms. It is impossible to travel in them without all the paraphernalia of a polar expedition, and very difficult to penetrate them from either side. A thin, jagged column, they stand right at the edge of the continent. Their western foothills and valleys are half submerged in the Pacific, forming an intricate maze of tiny islands, needle-sharp inlets and twisting channels. On the eastern side, the only way to enter them is to trek up the side of a glacier.

From the point I had reached in my flight, the Andes stretch south, losing some of their height but none of their hostility, until, briefly interrupted by the Strait of Magellan, they run down into the archipelago of Tierra del Fuego and finally peter out in a group of islets at Cape Horn, the storm-blasted extremity of South America. This is the nearest any continent comes to Antarctica. Tierra del Fuego was christened "the uttermost part of the earth" by E. Lucas Bridges, the son of a British missionary, who lived there from his birth in 1874. Bridges took his text from that splendid passage in the first chapter of Acts: "And ye shall be witnesses unto me both in Jerusalem ... and unto the uttermost part of the earth." The name was well chosen and I have always associated it in my own mind with the whole of the southern Andes, both in Tierra del Fuego and in Patagonia.

Patagonia was named in the 16th Century by Magellan. As he sailed down the Atlantic coast of what is now southern Argentina with his fleet of four ships, searching for a passage to the Pacific, he discovered the Tehuelche Indians. Either because they had large feet, or because of the bulky skin coverings they wore as shoes, Magellan called them Patagones, or "big-feet", and the name was later applied to the whole area.

I had already seen Patagonia in summer while visiting the eastern foothills of the Andes. Their forested slopes were decked in deep green and orange tints that gave them the beauty of northern woods in autumn. Now I was to make my first winter visit. From the air it was clear that the country at the foot of the mountains was thickly covered in snow and ice, and travelling would be difficult. But I was determined to get to the spectacular Moreno Glacier.

Periodically, the glacier advances across a narrow section of Lake Argentino, forming a giant dam and ploughing into the forest on the far shore. If I could watch the glacier creeping forward in this unique manner, I felt I would be transporting myself back to the Ice Ages, when glaciers advanced throughout the temperate zones, flattening the primeval forests and destroying the prehistoric animals or driving them back into the safety of the tropics.

The Ice Ages were predictably severe in Patagonia because of its closeness to the South Pole. At the height of each glaciation, the ice sheets were probably more than 4,000 feet thick on the Pacific side of the Andes, and their great weight depressed the coastline. Giant glaciers carved into the young mountains, giving them their characteristically sharp profile. The glaciers flowed down into the sea on the west, and on the east carried debris far into the Patagonian plain. Darwin found numerous boulders isolated on the *pampas*, and correctly described them as "erratics": rocks left behind when the glaciers retreated.

The Patagonian ice-cap is only a small remnant of these gigantic ice sheets. Nevertheless it covers 9,000 square miles and its two sections almost completely fill a 300-mile trough between the twin cordilleras. Moreover, it still demonstrates the processes that caused the ice sheets to expand. In this southerly latitude, the winds over the Andes blow from the west, bringing in moisture-laden air from the Pacific. Much of this moisture is lost as rain over the Chilean coast, and nearly all the rest falls over, and freezes on, the ice-cap. In this way the ice-cap is liberally replenished, and so great is the weight of ice that it is extruded outwards, creating glaciers like the Moreno which flow down the valleys. The glaciers stop only when they reach a lower, warmer level at which the rate of advance is equalled by the average rate of melting.

The 200-feet-thick Moreno Glacier advances across an arm of Lake Argentino. Periodically, it grinds into southern beech trees in the foreground.

I started my journey into the Patagonian Andes from Río Gallegos, a small town on the Atlantic coast. Marion was unable to come with me on this occasion and I was to join one of the special flights made by the Argentinian armed forces to remote parts of the country. My destination was Calafate, a small settlement on the shores of Lake Argentino, at the base of the mountains. Calafate is the starting point for any visit to the Glaciers National Park, which encloses a large area of forested foothills and, among several others, the Moreno Glacier. In the brief summer, when the temperature is well above freezing and the days are long, the park receives many visitors, but in mid-winter I was the only one heading west.

I had risen in darkness. As I walked to the waiting aircraft, the fiery glow of a magnificent, cold dawn streaked the sky. The flight was smooth and for nearly an hour we maintained a steady north-westerly course, never much more than 3,000 feet above the Patagonian plain. The Andes formed a line of snow-covered peaks to the west. Farm roads and fencing passed below. From time to time I saw a solitary vehicle and an occasional small homestead, but for the greater part of the journey the land seemed empty. The soil, always sparsely vegetated, has been eroded by wind, rain and by sheep farming. And during the summer the fierce *pampero*, the incessant wind of the plains, raises dust storms that travel hundreds of miles to the Atlantic. It is said that if you wish to see Patagonia in this season, you need not take a step—just stand still and let Patagonia blow past you.

The pilot put the aeroplane down on the airstrip at Calafate, with the Andes towering behind us. I went to the log-cabin headquarters of the park administration, and told the officials that I wanted to reach the Moreno Glacier. They looked at me glumly and shook their heads: "Impossible, señor!" They turned away to warm their hands at a coal fire. Although the winter was mild by local standards, the temperature was well below freezing in the mountains, and the track to the glaciers was covered in ice. No vehicle could hope to get through. You can also make the journey in a launch run by the park administration, but this boat was being overhauled. No one wants to go to the glaciers in winter.

I suggested driving as far as possible and then continuing on foot, and this idea met with grudging enthusiasm. "Yes, why not try Pedro Ojeda? He is a good driver and may get through in his *camioneta* (a small pick-up truck)." I tried Pedro Ojeda, and he agreed to make an attempt. There was a heavy cloud in the mountains to the north-west as we set off.

Pedro warned me that if it snowed again, there would certainly be no chance of getting more than a mile or two down the road.

Our route lay along the southern edge of Lake Argentino. On the left, the vegetation was sparse and dry. The few trees and bushes were stunted and twisted by the wind. Within half an hour we were appreciably nearer the mountains. Somewhere ahead of us where the farther reaches of the lake curved round sharply to the left was the Moreno Glacier. Through a wide opening in the hills I could see a distant range of chiselled mountains in the south. "Que suerte!" exclaimed Pedro with some justification. What luck! They were the main peaks of the Cordillera del Paine, the unmistakable landmark of Ultima Esperanza Sound on the other side of the Andes. The Paine are normally obscured by the storms that come in from the Pacific, and it was exceptional to be able to see them from here.

We turned south-west away from the lake and began to climb gently around the base of a low hill, a spur of the Sierra Buenos Aires. Before we had gone a mile, we came to a stretch of road covered with a glassy sheet of ice. Water from a stream in the *sierra* had frozen across our path. Pedro approached it carefully. The tyres slipped only slightly, and we managed to get across. "There will be more like that," he said. "The ice sheets get thicker, and the road becomes steeper soon."

By now we had entered the thick forest of the foothills. The trees, species of southern or austral beeches, were bare of leaves, their trunks were dark and the place was sombre. Snow clouds sweeping over the mountains ahead deepened the gloom. We could no longer see the main body of Lake Argentino, and on our left a steep slope led down to the edge of the Brazo Rico, an upper arm of the lake, 13 miles long. A few trees, grey and apparently dead, stood on the shore. In his soft Patagonian dialect, so different from the almost guttural Spanish I was familiar with, Pedro explained that there had been a huge flood which had been caused by the Moreno Glacier.

Pedro had been right about the ice. We encountered one patch after another and soon the road began to disappear beneath it. Often only the line of trees and the edge of the cliff marked the route. The slightest rise or dip was a challenge, and it was clearly foolhardy to drive on. With the glacier still some miles away, Pedro suggested that we continue on foot. We chocked the vehicle securely with broken branches. Stones were abundant but so firmly frozen into the ice that we could not prise them free, even with Pedro's pickaxe.

As we started walking, it began to snow. To avoid some of the twists in the road, we took a short-cut along some trails that ran over the lower spurs of the *sierra*. The ground was frozen hard and when I stamped my feet it rang as though hollow. Some streams were still running in the gullies, but they flowed under a crust of ice so thick that there was no danger of falling through. The stream-side boulders and plants were encased in glassy sheaths of ice. There were many fallen tree trunks, blanketed with soft dry snow. When I brushed some of it away I exposed a heavy covering of dark green moss. Some trees were festooned with manes of a yellowish-orange parasite, *Myzodendron punctulatum*, a flowering plant of the same order as the mistletoe.

A golden mass of Myzodendron punctulatum, a semi-parasitic shrub similar to mistletoe, brings a burst of colour to a Patagonian beech forest in the depth of winter. The plant puts out roots that penetrate its host tree. Although the myzodendron is able to manufacture some of its food by photosynthesis, it relies on the tree's sap for water and minerals.

The forest seemed totally empty of animal life, and I was beginning to think that none could survive. But when we stopped to consider our route, a hare ran out, stopped a yard or two away and stared at us. For almost half a minute it remained still, as if astonished to see us. It was one of the hares introduced from Europe that now plague parts of Patagonia. Hunger and the scarcity of people had made it tame. Just as we became impatient from standing still, the hare loped away.

We moved on. Suddenly the silence was shattered by a loud chattering. I looked up and saw a group of more than a dozen reddish-green parrots perched on the bare branch of a tree. If birds can look miserable, these did. Their feathers were ruffled against the cold, and when I approached them, they did not stir. They were *cotorras*, or austral parakeets, and their unexpected presence reminded me that another rare bird is found in the Andean forests as far south as Tierra del Fuego. It is the Patagonian firecrown, a species of hummingbird which is believed to go into a state of torpor, or temporary hibernation, during very cold periods. No doubt there were some of these birds in the forest too, huddled and motionless on some icy branch, but I did not find them.

Every step seemed to take us into thicker ice. It was as though the entire forest were growing inside a freezer. Close to the ground, the bushes were covered with needle-like ice crystals, formed when the mountains had been swathed in the freezing *niebla* of low-lying cloud. The ice crystals had later been dusted with snow. When I snapped a twig from one bush, I could see that a thick, tough bark protected its greenish living core from the ice.

It was now late afternoon, and the sky, already darkened with snow, seemed heavier and more forbidding than any I had seen in the Andes. Above the dark line of trees ahead, black rocks were exposed. We reached the top of the rise, and then I had my first sight of the Moreno

Glacier. It was ample reward for my long and hazardous journey.

At the bottom of the hill was the Canal de los Témpanos, or Channel of the Icebergs, which connects the Brazo Rico with the main body of the lake, now behind us to the right. On the far side of the channel was a wide valley leading up into the mountains. This entire valley was filled by the glacier, a thick carpet of jumbled ice that spread down into the Canal de los Témpanos and reached across it—a great white promontory advancing into the lake.

The crevassed and fluted face of the ice towered 200 feet above the surface of the water, glowing with a pale phosphorescence in the grey winter light. At intervals, huge chunks of ice fell from it, hundreds of cubic yards at a time. Sometimes a sharp, explosive crack preceded the collapse, and sometimes the ice fell noiselessly until it struck the water. Often the splash was as high as the glacier itself and as the massive white blocks plunged one after another, disappearing momentarily beneath the surface, they sent waves surging towards the shore.

Behind the Moreno's chill blue face, 27 square miles of ice flows down from the Andes, constantly impelled by the huge growth of ice in the Patagonian ice-cap 4,300 feet above sea level. Since records were started in 1899, the glacier has made repeated advances across the Canal de los Témpanos, moving at up to 500 yards a year. In 1939 it dammed the channel completely. It has done so on a number of occasions since. In 1970 it not only dammed the channel, but also ploughed into the forest on the shore, turning the trees to matchwood and scouring the rock slope clean. Behind the dam, the water of the Brazo Rico rose slowly until, after two years, it reached 123 feet above its normal level. Thousands of acres of land around the Brazo Rico were flooded, including the steep shoreline I had seen from the *camioneta*.

In March 1972 a section of the ice wall began to glow intensely blue and a giant fracture appeared. The glacier broke up, and thousands of tons of ice exploded outwards. An archway 190 yards wide and 132 feet high was carved out, and the trapped water of the Brazo Rico poured through at an estimated 8,000 cubic metres a second. For two days the flow continued unabated, and it was only after some weeks that the two lakes equalized, leaving the main part of Lake Argentino seven feet higher than normal. Some idea of the volume of water can be gained when you consider that this section of the lake covers an area of 500 square miles.

At the time I saw the glacier, the excess water had drained away down the river Santa Cruz into the Atlantic Ocean and the snout of the glacier lay a good hundred yards from the shore. The glacier itself stretched up like a highway into the mountains, and seemed to beckon me into the ice-cap. Indeed it was here that the first climbers ascended to the ice-cap. They were a team sponsored by the Argentinian Scientific Society and led by Frederick Reichert, who continued exploring the Patagonian Andes into his 60s. The party began their journey in 1914—very late in the history of exploration, considering that both Scott and Amundsen had already reached the South Pole. Because the surface of the glacier is broken into countless ice columns formed by partial melting during the summer, they travelled up the moraine on the glacier's northern side.

I strained to catch a glimpse of the moraine, but a biting wind hurled a blizzard of snow and ice particles through the gap in the mountains, stinging my face and forcing me to look away. I turned regretfully, for I was awed and fascinated by standing so close to one of the world's least explored places.

When Reichert and his team reached the ice-cap, they found a level plain of snow stretching into the distance, broken only by treacherous crevasses. They travelled for two weeks, and on calm days made good progress. But frequently, for days on end, the westerly winds howled in from the Pacific at velocities of more than 100 miles an hour, driving ice particles into the explorers' faces and sapping their strength. Rain and snow soaked through their tents, and even when the winds dropped, the sub-zero temperatures made life a grim struggle.

Reichert got only as far as the centre of the ice-cap, where a range of mountains up to 11,000 feet high forms a central divide and bars the way to the west. But the lure of the unknown drew him back on many subsequent journeys, and by the 1930s he and another explorer, a Salesian priest, Padre de Agostini, had explored much of the eastern part of the ice-cap. De Agostini wrote exhaustively about the geography of this part of the ice-cap and the surrounding Patagonian regions, and captured the grandeur of the mountains in panoramic photographs.

Only a handful of other men have travelled on the ice-cap, all of them since the Second World War. Among the most persistent are two Argentinians, Emilio Huerta, a short, tough colonel from the armed forces, and Mario Bertone, a glaciologist. With three companions, they made the first complete crossing of the ice-cap in 1952. Since then they have established a small base on the ice, supplied by air in the Antarctic manner. The British explorer, Eric Shipton, a veteran of six Everest climbs, has made three expeditions, including the crossing from west to east. This was also the longest journey, for in the course of

Heavily eroded rock spires including FitzRoy (centre) and Torre (left of centre) rise behind a glacial lake on the Patagonian ice-cap.

52 days Shipton travelled the length of the southern part of the ice-cap.

The uncertainties of the ice-cap are well illustrated by the mystery of an elusive volcano. It had been clear for a long time that Cerro FitzRoy, although named Chalten by the Indians, was not in fact a volcano. But the Indians reported that they had found ash on the Patagonian plain, blown from the mountains, so it seemed likely that there was an active volcano somewhere. The first definite evidence came in 1933 when Reichert caught a glimpse of a volcano during a lull in a storm. But he was not able to pinpoint it. Later, signs of volcanic activity were detected from aerial photographs but it was not until 1960 that anything definite was found.

In that year, Shipton mounted his second expedition, with the intention of searching for the volcano. Ascending the O'Higgins Glacier close to Cerro FitzRoy, he trekked far into the ice-cap, hauling sledges in strong winds and driving rain. One evening the tempestuous weather cleared, and in the west Shipton saw a long, impressive line of peaks silhouetted against the sunset. The highest mountain had a black fissure on its icy slopes from which there flowed a steady stream of vapour. This was undoubtedly the volcano Reichert had seen 27 years before. It was a lucky chance that the volcano was showing outward signs of activity at the very time when Shipton was there to see it and lucky indeed that the weather had cleared at the right moment. Shipton was now able to provide a precise location for the mountain, Cerro Lautaro, and in 1973 a British climber, Leo Dickinson, and two companions climbed it, also identifying two neighbouring mountains as volcanoes.

Nearly all the explorers of the ice-cap started and finished their journeys on the eastern side, for the stormy, rain-soaked Pacific coastline is difficult to reach. Like them, I had little chance of getting down on the other side of the ice-cap, but the same splintered conjunction of land and sea continues right through Tierra del Fuego, and I hoped to explore some of it there. After leaving the Moreno Glacier, I returned to Calafate and so to Río Gallegos. From there I took another flight down to Ushuaia, one of the world's most southerly towns.

Like Patagonia, Tierra del Fuego is divided into two distinct regions: flat and dry on the east, mountainous, wild and rainswept on the west. It is cut off from the rest of the continent by the Strait of Magellan, which narrows in the west where it passes through the Andean backbone. South of the channel on the horizon lies the snowy, triangular shape of Mount Sarmiento, and beyond Sarmiento the Cordillera Darwin leads the Andes southwards to Cape Horn.

In its western and southern parts, Tierra del Fuego disintegrates into a maze of waterways, including a second east-west route, the Beagle Channel. In many of these inlets and canals, glaciers flow into the sea. Charles Darwin, who spent two years in this region between 1832 and 1834, recorded his astonishment when he first saw a range only 3,000 to 4,000 feet in height, "with every valley filled with streams of ice descending to the sea-coast. Great masses of ice frequently fall from these icy cliffs, and the crash reverberates like the broadside of a man of war through the lonely channels."

These icy waters became the graveyard for generations of sailors rounding the Horn to the Pacific, to the goldfields of California, or later to the nitrate fields of Chile and the guano beds of Peru. Even today the shores of Tierra del Fuego are littered with the hulks of sailing ships, victims of the notorious westerly winds, the roaring forties. Many of the names left over from those days record nothing but frustration and despair: Useless Bay, Desolation Island, Port Famine, Cape Deceit.

Few people visit the outlying part of Tierra del Fuego; its mountains have been explored even less than those of Patagonia. One of the first was Darwin, and it is his descriptions that for me sum up the character of this wild region. In the 1830s, Tierra del Fuego was virtually unknown to the outside world, and Darwin's reports betray all the candour of a man of 23 describing to civilization an utterly primitive place. The dense forests reminded Darwin of the tropics, but there was a difference, for here they were gloomy, wet and cold. "In these still solitudes," wrote Darwin, "Death, instead of Life, seemed the predominant spirit."

During many of his journeys in this region, Darwin travelled in a small cutter. Along the bleak coastal inlets and channels, he came into frequent contact with the indigenous people, nomadic Canoe Indians known as the Alacaluf and the Yahgan. The western parts of Tierra del Fuego and the Pacific channels of Patagonia were the home of the Alacaluf; and the Yahgan, the southernmost inhabitants of the world, lived around the coasts of the Fuegian islands as far south as Cape Horn. It was a source of amazement to Darwin, and to other occasional visitors, that they were able to make a living in this god-forsaken place.

They dived into the icy sea to collect shellfish, sea urchins and crabs, and they gathered wild plants, berries and seeds from the forests. One other tribe, the Ona, a people of the Fuegian plains, hunted the guanaco, a relative of the vicuña. To all these Indians, fire was of paramount importance. They kept fires constantly alight wherever they went, and the Yahgan and Alacaluf even carried burning embers in their canoes. It was this custom that prompted Magellan to give the archipelago its name, Land of Fire.

The most primitive Fuegians were the Yahgan who, in spite of the extreme harshness of the climate, wore nothing more than scraps of animal skin in the manner of capes. Some wore nothing at all. On one occasion near Wollaston Island, less than 25 nautical miles from Cape Horn, Darwin came upon a canoe carrying six Yahgan Indians whom he described as "the most abject and miserable creatures I anywhere beheld". They were quite naked, and chilling rain ran down their bodies. In a bay not far away, an Indian woman came alongside his cutter suckling a recently born child "and remained there out of mere curiosity while the sleet fell and thawed on her naked bosom, and on the skin of her naked baby".

Darwin was clearly shocked by what he had seen. "It is a common subject of conjecture," he remarked, "what pleasure in life some of the lower animals can enjoy; how much more reasonably the same question may be asked with respect to these barbarians! At night five or six human beings, naked, and scarcely protected from the wind and rain of this tempestuous climate, sleep on the wet ground, coiled up like animals. . . . If a seal is killed, or the floating carcass of a putrid whale discovered, it is a feast; and such a miserable food is assisted by a few tasteless berries and fungi."

The question of how the Fuegians resisted the cold was another subject for conjecture. Darwin believed that they were in some way adapted for the purpose, and he was later proved right, for it was discovered that they had a higher metabolic rate than the human average. However, there are some more straightforward explanations. Early visitors, including Darwin, noticed that the Indians had bloodshot eyes, and this may have been an indication that they spent much of their time sitting over their fires. Moreover, their lack of clothes—at first sight inexplicable—had one distinct advantage: their bodies became wet, but dried quickly. Clothes would have remained damp much longer and, in fact, when they were encouraged by missionaries to adopt European clothing later in the 19th Century, the Indians appeared to become much more susceptible to cold.

The Indians no longer live their traditional life in the wild coastal regions. They have been almost entirely absorbed into the general, mixed-blood population of southern Chile. Although the Chilean government has established new settlers, I imagined that I would find

A rare, wild guanaco, close relative of the vicuña and the llama, peers through a clearing in a lichen-draped southern beech forest.

Tierra del Fuego even more deserted and desolate than it had been in Darwin's time. Moreover, Darwin had written about the forests only in summer; and if they were gloomy then, I wondered what they would be like now, in the depth of winter.

Ushuaia, my first objective, can be completely cut off from the outside world, for the road leading to it from the north passes over high, snow-swept mountains and the airstrip is often closed by ice. It was closed on this occasion, and I had to wait for some days at Río Grande, 70 miles to the north, before completing my flight. No aircraft had been into Ushuaia for ten days. Finally the weather cleared and I was able to fly in, but I was warned to take the first flight out if the weather closed in again. Otherwise there was no telling how long I might be stranded.

Ushuaia stands on the northern shore of the Beagle Channel and since many of its buildings are of wood, it has the air of a frontier town. It is a common jibe that there is running hot and cold water in Ushuaia: hot in summer, cold in winter. It certainly was a desolate place. The days were short and dark and the western sky seemed to be perpetually overcast.

Only one road leads west from Ushuaia. It runs along the northern side of the Beagle Channel, and passes through a wild stretch of Fuegian forest. Beyond this lies the Cordillera Darwin, the last major range of the Andes, an impenetrable wilderness of forest and snow. I borrowed a small truck and started out along the road. After my experience in Calafate, I was prepared for the ice and when I had driven as far as I dared to go, I ventured on foot into the wild forested slope leading down to the Beagle Channel.

As I entered the forest I could see few of the similarities with a tropical forest mentioned by Darwin. In the summer month of January, when Darwin did most of his land exploration, many plants would have been in flower and the trees would have been richly coloured. Today everything was grey or black. The ground was frozen hard and much of it was snow-covered. I moved down the slope cautiously, but there were so many plants under the blanket of snow, close to the ground, that I was able to get a good foothold. Where streams had flowed earlier in the year, the ground was thick with ice—glassy cascades that held small plants like relics preserved from the past. In one place I brushed the snow aside to see what I was standing on. It was ice, and beneath it the floor of the forest was green.

Although the hidden plants provided me with such a firm foothold, my progress was very slow. Fallen trees and branches lay everywhere, an obstacle course horribly embellished with ice and snow. In this cold climate, dead wood is slow to rot, and over the years it accumulates, giving the forest a tumbledown appearance. I had never seen anything so sombre and I could understand what Darwin had meant when he wrote about the spirit of death.

Nevertheless, as I continued down the slope, I found further signs of life. Some of the largest trees still bore leaves. They were the *guindos*, the evergreen southern beeches characteristic of Fuegian forests and known as *Nothofagus betuloides*. They were well over 60 feet high. I found two other species of *Nothofagus*, including the *Lenga* (*Nothofagus pumilio*) whose leaves turn a deep red in summer and autumn, and the *ñire* or *Nothofagus antarctica*. I also saw the Winter's bark tree, or *canelo*, whose bark was used by early navigators to relieve scurvy. In the undergrowth were spiny-leaved barberries and the straggly branches of Magellan's currant (*Ribes magellanicum*). Neither of these had any fruit, but I discovered a low, woody shrub half buried in snow which, in spite of the harshness of winter, was covered with small red fruits. It was *chaura*, and its edible fruits once provided the Indians with part of their winter diet. I soon found that the plant was abundant, and that there were other sources of food in the forest. On some trees hung clusters of orange, globe-like fungi. To Darwin they had seemed like tiny tangerines, and he described the species in some detail. "When young, it is elastic and turgid, with a smooth surface; but when mature it shrinks, becomes tougher, and has its entire surface deeply pitted or honeycombed. In Tierra del Fuego the fungus in its tough and mature state is collected in large quantities by the women and children and is eaten uncooked. It has a mucilaginous, slightly sweet taste, with a faint smell like that of a mushroom." My fungi were pale and tasted unpleasant, so clearly they were not mature.

I found a way that led down to the sea. Cold grey waves beat solemnly against the rocky shore, flecked with white from the flurries of snow that blew down the channel. I muffled myself against the icy wind. Somewhere to the west, hidden by heavy snow clouds, lay the Cordillera Darwin, guarded by barricades of forests and glaciers, and by the wrath of the savage climate. Only a hundred miles to the south was Cape Horn. I had come to the uttermost part of the earth, and the end of the Andes.

Patterns in the Andean Deserts

In the tropical north, the Andes are smothered in greenery; in the far south they are sheathed in snow and ice; but in the central ranges of Peru, Bolivia and Chile, desert has claimed much of the mountain area. Deprived of cover, the mountains stand here in all their majesty, as they were shaped by eons of geological and climatic forces, and with their textures and colours—red, ochre, purple, white—laid bare.

Now, although rain seldom falls in the desert, the ancient marks of water are visible everywhere in surfaces that were fretted into abstract patterns by prehistoric downpours, and in serpentine gullies that once frothed with torrents. The sun that now mercilessly bakes rock and soil, etches them with deep shadows. And across this parched realm blows the sculpturing wind. It hones the sharp ridges of the Chilean cordilleras; licks the brine of the Bolivian Laguna Colorada into baroque white coils; uncovers the delicate grain of rock strata, and heaps sand upon sand in the enormous dunes of the Peruvian coast.

KNIFE-EDGED RIDGES OF THE CHILEAN CORDILLERA DE LA SAL

RIVERINE GULLIES IN WATERLESS ANDEAN FOOTHILLS

AN IMMENSE WIND-RIPPLED SAND DUNE ABOVE THE PERUVIAN COAST

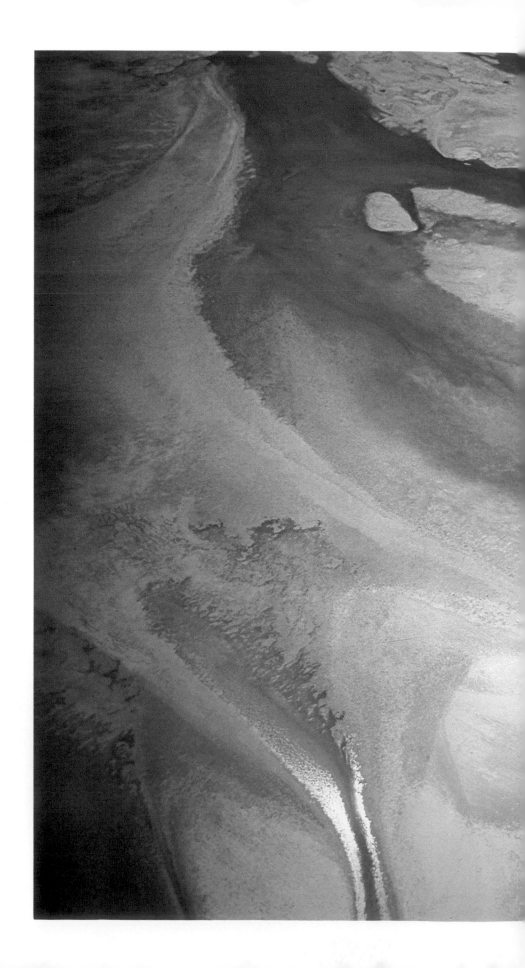

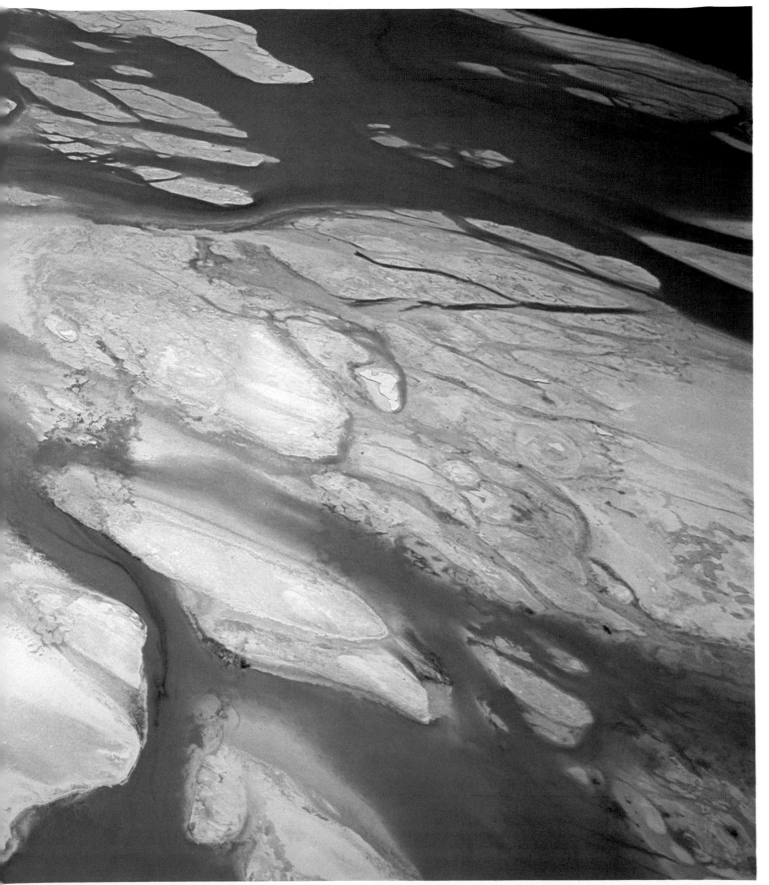

SALT-STREAKED SURFACE OF THE LAGUNA COLORADA

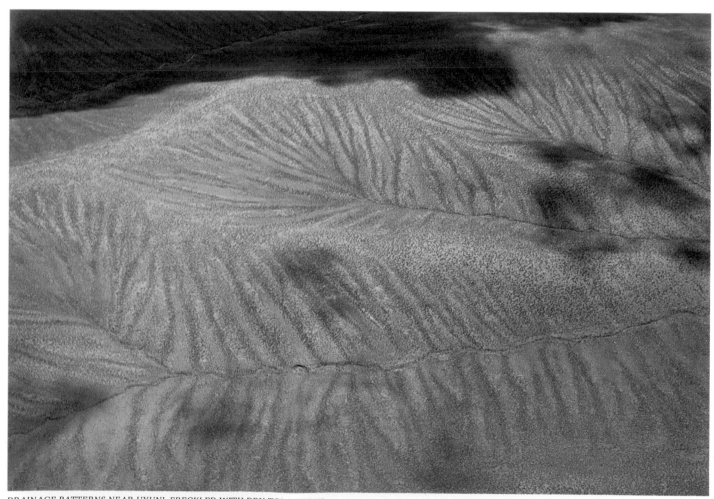

DRAINAGE PATTERNS NEAR UYUNI, FRECKLED WITH DRY TOLA SCRUB

DELICATE ONION-RING PATTERNS OF SAND, CUT BY PERUVIAN WINDS

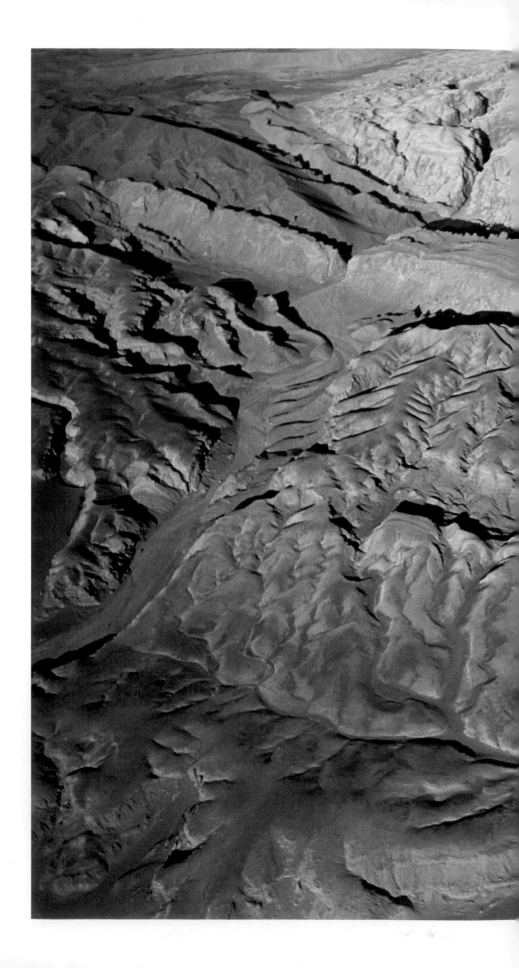

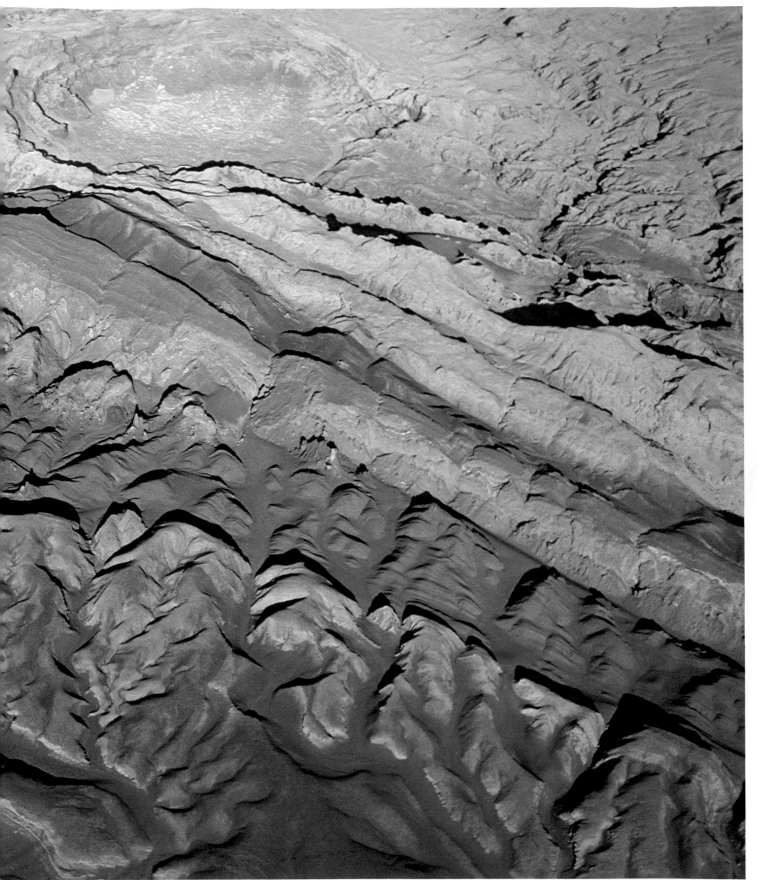

SUN-SCORCHED HILLS AND OUTCROPS

Bibliography

Aarons, John and Vita-Finzi, Claudio, *The Useless Land.* Robert Hale Ltd., 1960.

De Agostini, Alberto M., *Andes Patagónicos.* Published by the author, 1945.

Ahlfeld, Federico E., *Geología Fisica de Bolivia.* Los Amigos del Libro, 1972.

Alegria, Ciro, *The Golden Serpent.* The New American Library of Canada, 1963.

Anstee, Margaret Joan, *Gate of the Sun; A Prospect of Bolivia.* Longman, 1970.

Botting, Douglas, *Humbolt and the Cosmos.* Michael Joseph, 1973.

Bridges, E. L., *Uttermost Part of the Earth.* Hodder & Stoughton, 1948.

Cabrera, Angel, *Catálogo de los Mamíferos de America del Sur.* Vols. 1 and 2. Revista del Museo Argentino de Ciencias Naturales, 1957.

Chapman, F. M., *Distribution of the Birdlife in Ecuador.* Bulletin of the American Museum of Natural History Vol. 55, 1926.

Conway, Sir Martin, *Climbing in the Bolivian Andes.* Harper, 1901.

Conway, Sir Martin, *Aconcagua and Tierra del Fuego.* Cassell and Company Ltd., 1902.

Darlington, P. J., *Zoogeography.* Wiley, 1957.

Darwin, Charles, *The Voyage of the Beagle.* Nelson, 1914.

D'Orbigny, Alcides, *Voyage dans L'Amerique Meridionale.* Chez Pitois-Levrault et Cie, 1835-47.

Dorst, Jean, *South America and Central America. A Natural History.* Hamish Hamilton, 1967.

Fawcett, Col. P. H., *Exploration Fawcett.* Hutchinson, 1953.

Fittkau, E. J., Illies, J., Klinge, H., Schwabe, G. H., Sioli, H., *Biogeography and Ecology in South America.* Dr. W. Junk, The Hague, 1968.

Foster, Mulford B., "Puya" in *National Geographic Magazine,* October 1950.

Goodman, Edward, *The Explorers of South America.* Macmillan, 1972.

Goodspeed, T. H., *Plant Hunters of the Andes.* University of California Press, 1961.

Gould, John, *Birds of South America.* Eyre Methuen, 1972.

Von Hagen, Victor Wolfgang, *South America Called Them.* Robert Hale Ltd., 1949.

Von Hagen, Victor Wolfgang, *South America, The Green World of Naturalists.* Eyre and Spottiswoode, 1951.

Hemming, John, *The Conquest of the Incas.* Macmillan, 1970.

Herring, Hubert, *A History of Latin America.* Heinemann, 1939.

Howell, Mark, *Journey Through A Forgotten Empire.* Geoffrey Bles, 1964.

Hudson, W. H., *Idle Days in Patagonia.* J. M. Dent and Sons Ltd., 1923.

Von Humboldt, Alexander, *View of Nature.* George Bell and Sons, 1878.

James, David E., "The Evolution of the Andes" in *Scientific American,* Vol. 229, No. 2. August 1973, pp. 60-69.

Johnson, A. W., *The Birds of Chile—and adjacent regions of Peru, Argentina and Bolivia.* Volumes 1 and 2. Platt Establecimientos Gráficos S.A., 1967.

Koford, C. B., "The Vicuña and the Puna" in *Ecological Monographs.* No. 27, 1957, pp. 153-219.

Mason, Alden J., *Ancient Civilisations of Peru.* Penguin Books, 1961.

Matthiessen, Peter, *The Cloud Forest.* Andre Deutsch, 1962.

Menard, A. W., "The Deep Ocean Floor" in *Readings from the Scientific American,* September, 1969. pp. 80-81, 92, 96, 97.

Moorehead, Alan, *Darwin and the Beagle.* Hamish Hamilton, 1969.

Morrison, Tony, *Land Above the Clouds.* Andre Deutsch, 1974.

Moser, Brian and Tayler, Donald, *The Cocaine Eaters.* Longmans, 1965.

Mutis, José Celestino, *Diario de Observaciones de José Celestino Mutis.* Editorial Minerva Ltda., 1958.

Osborne, Harold, *Indians of the Andes.* Routledge and Kegan Paul Ltd., 1952.

Philipson, Dr. W. R., *The Immaculate Forest.* Hutchinson, 1952.

Prescott, W. H., *History of the Conquest of Peru.* The New American Library of World Literature, 1961.

Ruíz, Hipólito, *Travels of Ruíz, Pavón, and Dombey in Peru and Chile (1777-1788).* Field Museum Press, 1940.

De Schauensee, Rodolphe Meyer, *A Guide to the Birds of South America.* Oliver and Boyd, 1971.

De Schauensee, Rodolphe Meyer, *The Birds of Colombia.* Livingstone, 1964.

The South American Handbook. Trade and Travel Publications, London, published annually.

Shipton, Eric, *Land of Tempest.* Hodder and Stoughton, 1963.

Steele, Robert, *Flowers for the King.* Duke University Press, 1964.

Steward, J. H., *Handbook of South American Indians.* U.S. Government Printing Office, 1948.

Vásquez de Espinoza, Antonio, *Compendium and Description of the West Indies.* Smithsonian Institution, 1942.

Verdoorn, Frans, *Plants and Plant Science in Latin America.* Chronica Botanica Company, 1945.

Whymper, Edward, *Travels amongst the Great Andes of the Equator.* John Murray, 1892.

Acknowledgements

The author and editors wish to thank the following: Carlos Lozado Acuña, Servicio Nacional de Parques, Calafate; Aerolíneas Argentinas; Nicholas Asheshov, Lima; Professor H. G. Barriga, Universidad Nacional, Bogotá; Felipe Benavides, O.B.E., World Wildlife Fund, Peru; Mario Bertone, Instituto de Hielo Continental Patagónico; Professor Leonard Branisa, Geobol, La Paz; Martin Brendell, London; Alec Bright, Museo del Oro, Bogotá; Tom Browne, London; Romeo Cafferata, Servicio Nacional de Parques, Calafate; Santiago Castroviejo, Madrid; Stephen Clissold, London; Christopher Cooper, London; Dr. P. J. Cribb, Royal Botanical Gardens, Kew; Charles Dettmer, Thames Ditton, Surrey; Hugo Echegeray, Quillabamba; Peter Ellison, London; Tony Holley de Gomez, Arequipa; Dr. R. Harley, Royal Botanical Gardens, Kew; Colonel Emilio Huerta, Instituto de Hielo Continental Patagónico; David R. Hunt, Royal Botanical Gardens, Kew; Instituto Geográfico Militar, La Paz; Dr. R. Kauffmann, University of Giessen, West Germany; Lida Von Schey Koromla, B.B.C., Buenos Aires; Alfredo La Plaza, La Paz; the late Dr. F. Carlos Lehmann, Museo de Historia Natural, Cali; Berta Levin, Servicio Nacional de Parques, Calafate; Linea Aerea de Estado, Argentina; Nuno Mancilla, Calafate; Dr. Gustavo Maldonado, INDERENA, Santa Marta; Ariel Martinez, INDERENA, Colombia; Dr. Hernando de Macedo, Museo de Historia Natural, Lima; Katherine McCrary, Buenos Aires; Loren McIntyre, Virginia; Brian Moser, London; Antonio Olivares, O.F.M., Universidad Nacional, Bogotá; Roberto Risso Patron, Argentina; Dr. Carlos Ponce del Prado, Dirección Forestal, Lima; Dr. Werner Rauh, University of Heidelberg; Dr. Franz Ressel, La Paz; Samper Ltda., Bogotá; Staff of La Biblioteca de Estudios Hispano Americanos, Seville; Staff of Canning House Library, London; Roy Steinbach, Cochabamba; Rod Usher, Weston, near Beccles, Suffolk; Colin Vardy, Natural History Museum, London; Dr. Dionisio Vilez, INDERENA, Barranquilla; Wendy and Peter Williams, La Paz; Piotr Zubrzycki, La Paz.

Picture Credits

Sources for the pictures in this book are shown below. Credits for the pictures from left to right are separated by commas; from top to bottom they are separated by dashes.

Cover–Loren McIntyre. Front end papers 1, 2–Loren McIntyre. Front end paper 3, page 1–Marion Morrison. 2, 3–Leo Dickinson. 4, 5–Loren McIntyre. 6, 7–Dr. M. P. Kahl. 8, 9–Loren McIntyre. 10, 11–Marion Morrison. 12, 13–Leo Dickinson. 18, 19–Map by Hunting Surveys Ltd., London. 23–Michael Freeman. 27–Loren McIntyre. 31–Michael Freeman. 32, 33–Marion Morrison from Keystone Press Agency Ltd., London. 34 to 37–Loren McIntyre. 38, 39–Leo Dickinson. 43–Loren McIntyre. 44–Courtesy of the Instituto de Cultura Hispanica, Madrid. 47–Marion Morrison, A. P. Smith-Marion Morrison. 50, 51–Loren McIntyre. 54 to 57–*Recueil d'observations de zoologie et d'anatomie comparée* by Alexander von Humboldt, 1799-1803. Eileen Tweedy, courtesy of The Zoological Society of London. 58, 59–*Florae Peruvianae et Chilensis* by Ruiz Lopez and Pavón, 1794. Eileen Tweedy, courtesy of the Director, Royal Botanic Gardens, Kew. 60, 61–*Voyage dans l'Amérique Méridionale* by Alcide d'Orbigny, 1839-42, courtesy of the Royal Geographical Society, London. 65–Hans D. Dossenbach from Natural Science Photos, London. 67–Map by Hunting Surveys Ltd., London. 68 to 71–Marion Morrison. 75 to 81–Loren McIntyre. 86–Marion Morrison. 88–Alfred Gregory. 89–Tony Morrison, Peter Williams–Peter Williams. 91–Tony Morrison. 92–Loren McIntyre. 96–Map by Hunting Surveys Ltd., London. 97, 98–Hans D. Dossenbach from Natural Science Photos, London. 99–Dick Philips. 100–Hans D. Dossenbach from Natural Science Photos, London. 101–Loren McIntyre. 102–Merlin D. Tuttle. 103–Tony Morrison. 104–Hans D. Dossenbach from Natural Science Photos, London. 105–Loren McIntyre. 111–Loren McIntyre. 112–Marion Morrison. 117 to 129–Marion Morrison. 132–Tony Morrison. 135–John Amatt–Dr. Edward S. Ross, Tony Morrison. 136–Tony Morrison. 139–Marion Morrison. 140 to 142–Loren McIntyre. 143–Dr. M. P. Kahl. 147, 148–Dr. Peter Francis. 149–Eric Shipton, F.R.G.S. 150 to 153–Dr. Peter Francis. 157–Loren McIntyre. 160–Tony Morrison. 163–Leo Dickinson. 167–Roger Perry. 171–Loren McIntyre. 172–Jacques Jangoux. 173 to 175–Loren McIntyre. 176–Dr. M. P. Kahl. 177 to 179–Loren McIntyre.

Index

Colour reproduction by Printing Developments International Ltd., Leeds, England—a Time Inc. subsidiary.
Filmsetting by C. E. Dawkins (Typesetters) Ltd., London, SE1 1UN.
Printed and bound in Italy by Arnoldo Mondadori, Verona. **XX**